2

AGRICULTURAL RESEARCH UPDATES

AGRICULTURAL RESEARCH UPDATES

VOLUME 28

Agricultural Research Updates

Additional books and e-books in this series can be found on Nova's website under the Series tab.

AGRICULTURAL RESEARCH UPDATES

AGRICULTURAL RESEARCH UPDATES

VOLUME 28

PRATHAMESH GORAWALA
AND
SRUSHTI MANDHATRI
EDITORS

nova
science publishers
New York

Library of Congress Cataloging-in-Publication Data

ISBN: 978-1-53616-253-0
ISSN: 2160-1739

Published by Nova Science Publishers, Inc. † New York

CONTENTS

PREFACE

In the opening study of Agricultural Research Updates. Volume 28, the capillarigation system's applications in vertical gardens were explored. The vertical garden model was designed for growing vegetables in home yards or within the vicinity of farming households, so they can have a regular supply of safe and nutritious food, especially during periods where water supply is scarce, excessive, or when maximizing the use of stored rainwater or kitchen wastewater.

Next, this compilation describes recent progress and findings on the physiology of petal growth in cut rose flowers. To increase the ornamental value of cut roses, it is desirable that the timeframe for complete flower opening be relatively slow; therefore, understanding the mechanism by which rose flowers open is necessary to improve flower quality.

The objective of the subsquent chapter is to present some recent findings about the influence of processing operations on the properties and bioactive compounds in carrots. To achieve this, different pre-treatments and processing operations are discussed, and their effects on some typical carrot products are described.

Additionally, the authors discuss theoretical aspects and recent developments in ready-to-eat spinach, with the goal of improving the preservation of nutritional quality after harvest and extendiung its shelf-life.

The authors also analyse 30 samples of fresh, frozen and IV range spinach, with the aim of identifying and quantifying nitrate and nitrite levels.

Following this, a study is presented on the scientific validation of the medicinal and edible utility of fruits of A. moschatus using standard protocols sin order to account for potential commercial utilization. About 15 nutritional and five anti-nutritional constituents of the fruits of A. moschatus were tested and compared with those found in A. esculentus.

In the penultimate chapter, the authors describe the current global status of conservation of cattle genetic resources and the cryo-technologies that are available to facilitate the protection of bovine genetic diversity. Worldwide adoption of such a policy has led to unique adaptations by cattle to a variety of local environments.

The final chapter summarizes the following research progresses: the structure and components of Nox complex, the functions of these components, and control strategies for B. cinerea.

Chapter 1 – Droughts and floods in the Philippines have become more intense and frequent, causing significant losses in crop production. To have alternative source of income in case rice fails due to these extreme events, rice farmers are usually advised to plant vegetables and other short duration crops. This led to the development of the *capillarigation* system in 2015, a do-it-yourself type and low cost variant of the drip irrigation system that makes use of capillary wicks to perform the function of the drippers. In this study, the system's application in vertical gardens (VG) was explored. The VG model was designed for growing vegetables right in the home yards or within close vicinity of farming households so they can have regular supply of safe and nutritious food especially during periods where water supply is scarce (drought), excessive (flood), or when maximizing the use of stored rainwater or kitchen wastewater. A self-watering VG garden was developed which can accommodate 320 plants (leafy vegetables) planted in pots that are attached to 16 hanging poles made from 12.5mm PVC pipes. Results of performance test showed that, with an occupied space of 1m x 1m, it can grow pechay (*Brassica rapa*) with an average marketable yield of 3.7 kg harvest for a growing period of

30 days with no significant difference on plant height as well as on the amount of water applied to individual pots regardless of their location within the vertical garden.

Chapter 2 – The rose (genus *Rosa*) is one of the msost important plants for ornamental cut flowers and use in ornamental gardens; therefore, understanding the mechanism by which rose flowers open would benefit both general plant science and commercial horticulture. Although many studies have focused on senescence to improve the vase life of cut flowers, there are few studies on the physiology of flower opening. The vase life of cut roses depends, at least in part, on the timing of flower opening. To increase the ornamental value of cut roses, it is desirable that the timeframe for complete flower opening be relatively slow; therefore, understanding the mechanism by which rose flowers open is necessary to improve flower quality. In this review, we describe recent progress and findings on the physiology of petal growth in cut rose flowers.

Chapter 3 – Carrots are rich in many bioactive molecules, which provide health benefits for the human body. Among these are carotenoids, vitamins, saccharides or phenolic compounds like anthocyanins and others. However, it is well known that when carrots are ingested fresh they provide the highest bioactivity, and the processing operations such as boiling, frying, freezing, drying, among others, induce some chemical and structural changes that may diminish the bioactivity of the compounds present. Furthermore, pre-treatment operations, like cutting or bleaching are also known to influence the quality of the food products. For example, the thermal pasteurization of carrot juice influences some vitamins and carotenoids. Also, the drying of carrots to incorporate into dried formulae greatly imparts the anthocyanins' content. Hence, the objective of the present chapter was to present some recent findings about the influence of processing operations on the properties and bioactive compounds in carrots. To achieve this purpose, different pre-treatments and processing operations are discussed as to their effects on some typical carrot products.

Chapter 4 – Spinach is an annual, green, leafy vegetable rich in health-promoting compounds, such as vitamins A and C, and flavonoids. For consumer's acceptance, spinach leaves must be green and turgid. However, spinach undergo fast senescence processes after ripening and the quality attributes are quickly lost during postharvest handling. Inadequate postharvest management can result in significant nutritional losses due to the fact these leafy vegetables have high-water activity, neutral to slightly acidic pH values, nutritional complexity and can sustain the growth of many species of bacteria, yeasts and molds. Spinach has relatively high respiration and water-loss rates, and is prone to tissue decay, microbial growth, and nutrients loss, all reducing the commercial potential, when not controlled. Spinach nutritional quality can be affected by postharvest technologies and storage conditions. Starting from spinach washing process until consumption, there is high potential for ascorbate and other nutrients loss, which are responsible for important bioactive properties such as antioxidant capacity. Recent postharvest technologies have been used to better maintain ready-to-eat spinach physiological and nutritional quality, namely the use of hydrogen peroxide, organ ic acids, gamma irradiation, UV, low irradiance pulses, ozone gas, intense pulsed light and a recently developed technology, the cold plasma, among others. In addition, packaging and storage conditions, focusing on temperature and relative humidity, are key factors affecting processed leafy vegetables quality, frequently used aiming to maintain high-value product quality throughout production, distribution and storage. The knowledge and control of postharvest technologies impact on ready-to-eat spinach allows the attempt to modulate nutritional quality, bioactive properties and health-important compounds content increasing commercial value. This chapter discusses the theoretical aspects and recent developments in ready-to-eat spinach to improve preservation of nutritional quality after harvest and extend its shelf-life.

Chapter 5 – Nitrates and nitrites are compounds that may accumulate in several foodstuffs, especially vegetables, due to their presence in the environment as consequence of the nitrogen cycle. Moreover, they are largely used in meat curing as food preservatives. It is well known that nitrates and nitrites may exercise some harmful effects on humans. So, the levels of these compounds in foodstuffs have to be monitored by health authorities, continuously. These monitoring activities have been intensified on vegetables during the last years, since a significant increase of residual level of nitrate, especially in leafy vegetables, has been verified. This increase seems especially due to the wide use of nitrogen fertilizers in agriculture. The accumulation of nitrate in vegetables depends on N-fertilizer application timing, temperature, water relation, cultivar and other variables. Moreover, some studies have underlined the simultaneous presence of not negligible amounts of nitrite. This last finding is significant for food safety, since nitrite is involved in the synthesis of carcinogenic nitrosamines, directly. The nitrite mechanism of synthesis in leafy vegetables is not clear. Presumably, it is a reduction product of endogenous nitrate. In this chapter, the results obtained by analysing 30 samples of fresh, frozen and IV range spinach, with the aim to identify/quantify the levels of nitrate and nitrite, are described. The analyses were carried out by using a validated/accredited analytical method by ion chromatography with conductivity detection. The most important findings have been compared with other available data worldwide, and with nitrate/nitrite levels detected in other types of vegetables. Very interesting results were obtained for both nitrate and nitrite. As expected, very high levels of nitrate were registered in spinach, where some concentrations higher than the European legal limit were verified. Concerning nitrite, detectable amounts of this contaminant were found, at levels up to 78.5 mg kg^{-1}. This study confirmed that spinach is among the most nitrate-contaminated vegetables and that control is needed. Moreover, taking into account the nitrite levels detected in some samples, it is possible suggesting the introduction, by health and food safety authorities, of specific limits for this compound in leafy vegetables.

Chapter 6 – Biodiversity prospecting is concerned with exploration, collection and screening of natural resources along with associated indigenous knowledge for their beneficial utilization. Fortunately, the Indian subcontinent is gifted with many food crops and genetically diverse populations of their wild relatives. The fruits of many of these wild plants are consumed by the rural folk and literature is replete with reports of their medicinal worth. Also, the fruits of some of these plants are rich sources of nutrients, vitamins, minerals and anti-oxidants. Therefore, it is imperative that such lesser known edible plants should be collected from the wild, documented, studied and conserved. *Abelmoschus moschatus* (L.) Medik, (Malvaceae), commonly called 'ambrette plant' or 'musk mallow' is a tropical semi-wild, perennial shrub native to India, and a close ally of *A. esculentus*, the common okra. The present study focused on the scientific validation of the medicinal and edible utility of fruits of *A. moschatus* using standard protocols so that it can account for potential commercial utilization. About 15 nutritional and five anti-nutritional constituents of the fruits of *A. moschatus* were tested and compared with that in *A. esculentus.* Just as in common okra, the carbohydrate and crude fibre content was relatively higher than the protein and lipid content. The vitamin content however, was higher than in common okra. The anti-nutrient content was comparable, except oxalates and phytates. The study also included the evaluation of the phytochemical constituents of the fruits. Seven of the eleven phytochemicals tested such as flavonoids, phenols, tannins, glycosides, cardiac glycosides, saponins and coumarins were present in fruits. Based on the results of the present investigation, it is evident that the fruits of *A. moschatus* can be included in the diet just as *A.esculentus* and the plant deserves due promotion to commercial cultivation.

Chapter 7 – Domestic animal genetic resources, including those of cattle, are important for food security, economic development and the cultural heritage of all countries. As with other livestock and poultry species, many cattle breeds are at risk of extinction. The principal underlying reason for the loss of both beef and dairy breeds is an escalating rate and scale of economic and urban development as a result of the increasing global human population. A direct effect of these associated factors is the great emphasis that is placed on agricultural sustainability and animal production systems to provide sufficient food for the ever-increasing number of humans even though there are decreasing natural resources for this food production. Government policies for cattle conservation have been developed on all continents but there are vast differences in detailed aspects of these policies among countries. Initiatives to conserve cattle genetics, including the provision of gene banks for cattle, are well established in industrialized nations. In contrast, less economically developed countries are currently experiencing problems in safeguarding the preservation of local cattle breeds although there exists an immense diversity of animal genetic resources. This disparity is due to differences in a range of key factors such as national strategies, economic status, engagement of private sector companies and available technological capacity. The successful oversight and management of cattle genetic resources has major implications for the biotechnology industry and thus for the growth of a nation's economy. The combination of applying cryobiology to germplasm conservation and implementing national strategies to maintain living genetic resources is, therefore, essential to ensure the sustainability and diversity of cattle breeds. To this end, the advancement of cryopreservation technologies and the international prioritization of conservation breeding are both imperative. In this chapter, we describe the current global status of conservation of cattle genetic resources and the cryo-technologies that are available to facilitate protection of this valuable natural reservoir of bovine genetic diversity. Worldwide adoption of such a policy has led to unique adaptations by cattle to a great variety of local environments.

Chapter 8 – Reactive oxygen species (ROS) play crucial roles in the defense and differentiation of multi-cellular organisms. Nicotinamide adenine dinucleotide phosphate (NADPH) oxidase complex serves as the major enzymatic producer of ROS in eukaryotic cell. The NADPH oxidase (Nox) complex has been found in many fungi, and its functions have been well studied in some plant pathogenic fungi. *Botrytis cinerea* is one of the most devastating plant pathogenic fungi, and represents a model organism to study the pathogenicity of aggressive fungal pathogens due to its broad host range and great economic importance. Much work has shown that Nox complex is involved in many biological processes of *B. cinerea*. Here, we will summarize the following research progresses: (i) the structure and components of Nox complex; (ii) functions of the components of Nox complex; (iii) control strategy for *B. cinerea*.

In: Agricultural Research Updates. Vol. 28 ISBN: 978-1-53616-253-0
Editors: P. Gorawala, S. Mandhatri

Chapter 1

THE *CAPILLARIGATION* SYSTEM: ITS APPLICATION IN VERTICAL GARDENS

Derose A. Sawey and Ricardo F. Orge*
Philippine Rice Research Institute, Maligaya
Science City of Muñoz, Nueva Ecija, Philippines

ABSTRACT

Droughts and floods in the Philippines have become more intense and frequent, causing significant losses in crop production. To have alternative source of income in case rice fails due to these extreme events, rice farmers are usually advised to plant vegetables and other short duration crops. This led to the development of the *capillarigation* system in 2015, a do-it-yourself type and low cost variant of the drip irrigation system that makes use of capillary wicks to perform the function of the drippers. In this study, the system's application in vertical gardens (VG) was explored. The VG model was designed for growing vegetables right in the home yards or within close vicinity of farming households so they can have regular supply of safe and nutritious food especially during periods where water supply is scarce (drought), excessive (flood), or when maximizing the use of stored rainwater or kitchen wastewater. A

* Corresponding Author's E-mail: rforge@gmail.com.

self-watering VG garden was developed which can accommodate 320 plants (leafy vegetables) planted in pots that are attached to 16 hanging poles made from 12.5mm PVC pipes. Results of performance test showed that, with an occupied space of 1m x 1m, it can grow pechay(*Brassica rapa*) with an average marketable yield of 3.7 kg harvest for a growing period of 30 days with no significant difference on plant height as well as on the amount of water applied to individual pots regardless of their location within the vertical garden.

Keywords: *capillarigation*, climate change, drought, flood, vertical garden

INTRODUCTION

Climate change is threatening the sustainability of the Philippine agriculture. Extreme droughts, for example, had become more frequent, causing significant losses in rice production. The World Resources Institute predicted that the Philippines, together with other Southeast Asian Countries, will experience a "high" degree of water shortage in the future. By 2025, 15 out of 75 million hectares of Asia's flood-irrigated rice crop will experience water shortage (Tuong and Bouman, 2003).

El Niño comes every 2 to 8 years (Vikas and Dwarakish, 2015) that resulted to rainfall reduction that may led to dry spells, drought and strong typhoons (FAO, 2017). Among the adaptation strategies recommended for the rice farmers is to use drought and submergence tolerant rice varieties and of improved crop varieties that are tolerant to pests and diseases. However, when the supply of water from the national irrigation system is reported to be in critical condition, it is already too risky to plant rice. Thus, in order to have an alternative source of income in case rice fails due to these extreme events, rice farmers are usually advised to, among other things, plant vegetables and other short duration crops through the *Palayamanan*, a highly diversified and integrated system of farming being promoted by PhilRice and its partner agencies. Under this system of farming, aside from planting rice, farmers are engaged in vegetable production as well as in other farming activities like duck raising (integrated in rice fields), mushroom production, poultry, livestock and aquaculture, etc) so as to diversify their sources of income (Corales et al., 2004) thus helping enhance their resilience to climate change (Brklacich et al. 1997, Smithers and Smit 1 997, de Loë et al. 1999, Kelly and Adger, 2000, Smit and Skinner, 2002).

To maximize the use of limited water supply, an efficient watering system is needed. The commercially available drip irrigation system was proven to be efficient, however, its cost of acquisition is high and beyond the financial capability of small holder Filipino farmers. This led to the development of the *capillarigation* system in 2015. This system of applying water to the plants is generally patterned from the drip irrigation system, but made use of capillary wicks (hence its name was taken from two words, 'capillarity' and 'irrigation') as means of dispensing water equivalent to the emitters in the drip irrigation system. The use of capillary wicks simplifies the design of the whole system making the fabrication of most parts easy and doable by the farmers themselves, resulting to a significant reduction in cost of up to 80%. The c*apillarigation* system was designed as a do-it-yourself type irrigation system to meet the requirements of smallholder farmers to help them cope up primarily with extreme drought conditions. More information about this technology was described in previous publications of the authors (Orge and Sawey, 2017).

The system can work with unfiltered and kitchen wastewater with very low operating pressure (15 – 20 cm water) and at discharge rate of 25 - 35 mL/h per wick. In terms of labor requirement, it requires minimal attention as there are no valves to open or close and filter to be cleaned as in the case of the drip irrigation system. Results from field tests conducted (Figure 1) showed that the *capillarigation* system outperformed the existing farmer's irrigation practices in terms of water productivity (Orge and Sawey, 2019).

While the *capillarigation* system was primarily designed as a technology to help farmers cope up during water scarce conditions, it also has the potential be used in vertical gardens during rainy season when it is not possible to plant vegetables directly on the ground which is often submerged in water. Moreover, the occurrence of rain during rainy season is becoming erratic hence there are also chances that plants are water stressed if purely depending on rainfall for irrigation. Vertical farming is also becoming popular in the urban areas due to issues on limited land area, food security and limited water supply (AbumhadI et al., 2012). There are other existing technologies like hydroponics, aquaponics and aeroponics associated with vertical gardens yet its establishment and operation requires high investment cost as well as high technical skills.

Figure 1. The *capillarigation* system as applied in a field planted with green pepper (Orge and Sawey, 2017).

This study explored the application of the *capillarigation* system in vertical gardens. Specifically, it aimed to develop a self-watering vertical garden model with built-in *capillarigation* system for growing green and leafy vegetables.

MATERIALS AND METHODS

This study was conducted at the Philippine Rice Research Institute – Central Experiment Station, Muñoz Science City, Nueva Ecija during wet season of 2018. During this period, the field used in the previous trials of the *capillarigation* system was submerged in water for least 2 days every after the occurrence of a heavy rain.

Design and Development of Self-Watering Vertical Garden Model

The vertical garden (VG) model developed in this study was designed for growing green leafy vegetables right in the home yards so farming households can have regular supply of safe and nutritious food especially during periods where water supply is scarce (drought), excessive (flood), or when maximizing the use of stored rainwater or kitchen wastewater. This allows saving on food expenditures of households. Also, it provides additional source of income through sale of excess harvest that cannot be consumed.

In this VG model, the *capillarigation* system was integrated as a means of applying water the plants that are grown in vertically-arranged pots to replace the manual watering of the individual pots.

Pechay (*Brassica rapa*) was used as test crop on the performance evaluation of the self-watering VG. A mixture of carbonized rice hull (CRH), garden soil and vermicast was used as growing media at 1:1:1 ratio by volume. Pechay was transplanted at eight days after sowing and at four plants per pot. The VG was covered with mosquito net to protect the plants from pests and birds. It was placed in such a way that it could be easily accessed during planting and harvesting.

In evaluating the performance of the system, this study focused only on gathering parameters for the watering system and on the growth and yield of pechay within the VG model. No application of fertilizer and pest control chemicals was done. In order to come up with replicated data for statistical analysis, additional units of the VG model were fabricated. Moreover, follow up cycles of planting using the same crop was also done in order to generate more data for the analysis.

Data Gathered

The following data were gathered to evaluate the performance of the system:

a) Water delivered per pot. Before planting, wick discharge rate per pot at different layers were taken to compare the uniformity of water distribution throughout the model. This was done for three trials.
b) Plant height. Measuring of this parameter was done on the largest plant within the sample pot. It was measured from the soil line to the tip of the longest leaf.
c) Marketable yield. After removing the roots and the unmarketable parts such as dried leaves, the remaining parts were measured.
d) Chlorophyll meter (SPAD) reading. This was measured using a chlorophyll meter to compare the chlorophyll content or greenness of the plants with reference to their location within the VG. Five readings were taken from the largest plant within the sample pot and the average was recorded.
e) Material and labor cost. The cost of materials of one model was computed including the labor cost for fabrication and assembly and installation of *capillarigation* system.
f) Labor requirement in terms of operation of the *capillarigation* system compared to that of manual watering or the farmer's practice.
g) Cost and return analysis. This was computed to determine the profitability of vertical garden model using *capillarigation* system.

Statistical Analysis

Whenever applicable, an analysis of variance (ANOVA) was done on the gathered data by considering the location of the plants within the VG model as treatments, following a Completely Randomized Design.

RESULTS AND DISCUSSION

The Prototype Vertical Garden with Built-In *Capillarigation* System

Occupying an area of 1m x 1m and with a total height of 2m, the developed VG model can accommodate 320 plants (pechay) planted in pots that are attached to 16 hanging poles made from 12.5 mm PVC pipes (Figure 2). The pots were made up of plastic mesh and one pot can accommodate four plants. The integration of the *capillarigation* system made the VG self-watering. This was made possible by establishing a steady supply of water at very low operating pressure through the use of a float valve-equipped container, positioned just beside or within the vicinity of the VG. From the water source (i.e household faucet), water enters into this container through the float valve and exits from another port going to the VG via the designated inlet port. Because of the float valve that controls the entry of water into the container, the level of water inside the container is maintained even if water is continuously flowing out from it. This container was attached to a ground-anchored pole in such a way that its vertical movement can be adjusted so as to maintain a certain level of water in the four horizontally-mounted GI pipes at the topmost portion of the VG which serve as a stable source of water for the plants. Each of these pipes supports four hanging poles and is equipped with four (corresponding to the number of poles) equally spaced 10mm holes that were drilled along the 1m pipe length. Capillary wicks (5-8 mm dia) made from cotton strands taken from mop heads (normally used for mopping floors) were inserted on each of the holes with the opposite end of the wick inserted on the top end of the hanging pole. Along the length of each hanging pole were five equally spaced 6mm holes drilled, corresponding to the number and the location of the pots. Additional wicks were also inserted on the holes, receiving drops of water being delivered by the topmost wick.

Figure 2. The VG model with built-in *capillarigation* system.

Water Application Rate

Figure 3 shows the location of the sample pots for data gathering that was done during harvesting. As presented in Table 1, the average water applied by the system to each pot ranged from 69 to 97 mL/hr. Numerically, the pot located at the lowest part received the highest amount of water although the difference is not statistically significant.

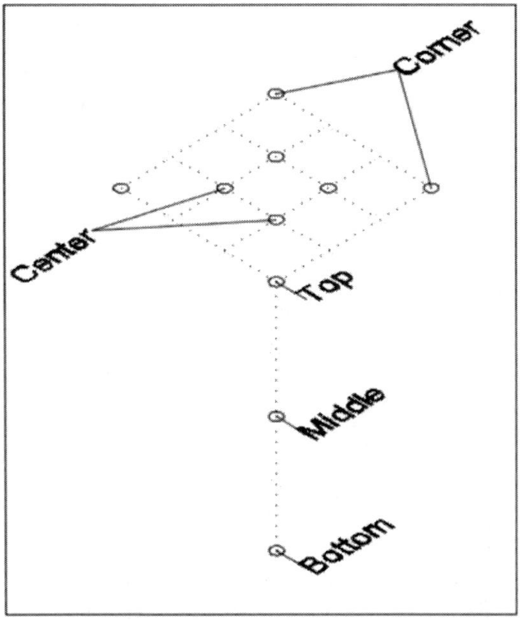

Figure 3. Schematic diagram describing the different locations of the pots in the VG.

Table 1. Average wick discharge rate at different pot locations

Location	Wick discharge rate per pot (mL/hour)*
First layer (top)	73
Second layer	78
Third layer (middle)	69
Fourth layer	86
Fifth layer (bottom)	97

* CV = 13.25%.

Crop Performance

As presented in Table 2, there was no significant difference on plant height, as well as on the greenness of the plants regardless of their location within the vertical garden. In terms of the total weight of marketable yield gathered per pot, no significant difference was observed except for the pots located at the corner (highest) and center (lowest) bottom. Further analysis done on per horizontal layer basis (i.e., from top layer down to the bottom layer) showed that the average values of these three parameters did not vary significantly. The same holds true when analyzed based on groupings in a vertical plane. From the results, the average marketable yield harvested after a 30-growing period was 3.7 kg and the average total number of plants harvested was 311.

Table 2. Performance test result of the self-watering VG based on relative location of individual pot

Location of sample pot	Marketable weight (g)*	Plant height (cm)	SPAD reading
Top corner	36.3^{ab}	15.1	30.3
Top center	51.3^{a}	15.5	33.9
Middle corner	40.5^{ab}	16.1	31.0
Middle center	36.2^{ab}	19.4	31.7
Bottom corner	53.4^{a}	16.2	30.0
Bottom center	23.4^{b}	19.3	31.4
CV (%)	17.7	19.0	10.8

* Means with the same letter does not vary significantly at 5% level of significance.

Material and Labor Cost

Table 3 shows the cost of materials and labor needed in fabricating the VG model with built-in *capillarigation* system. The total computed cost amounted to Php 5,400.00 of which 82% was for materials and the rest was for labor. Because it is self-watering, it does not require labor in watering the plants once installed. If done manually, plants grown in vertical gardens need to be watered twice a day or a total of 12 labor hours is required in a 30-day growing cycle.

Table 3. Material and labor cost for the fabrication of the components of the VG model

Component	Material used	Cost of component
Frame	7m GI pipe, 1" diameter	525.00
	8m GI pipe, 3/4" diameter	440.00
	4m GI pipe, 1/2" diameter	200.00
	5m GI pipe, 1-1/4" diameter	700.00
	Welding rod	25.00
Hanging pole	8 pieces PVC pipe, blue, 1/2" diameter	520.00
Plant pot	1.2m x 5m plastic mesh, green, 1/8" mesh	555.00
	Masking tape	20.00
Component for dispensing water	Cotton twine (from mophead)	50.00
Component for water source and control	1 piece float valve, brass, 1/2" diameter	480.00
	3m GI pipe, 1" diameter	225.00
	1 piece float valve tank	100.00
	4m flexible hose, 3/4" diameter	100.00
Component for protection against pests and birds	1.2m x 10m mosquito net	460.00
	Nylon string, 1mm diameter	30.00
Labor cost on the fabrication and assembly of the VG and installation of *capillarigation* system		970.00
Total		5,400.00

Cost and Return Analysis

Using the data taken from the conduct of this study (3.7 kg yield, 30 days growing cycle, preparation of seedlings done 8 days prior to harvesting, transplanting done 5 days after harvesting), it is feasible to grow pechay 10 times in a year using the VG model. Thus, for an average marketable yield of 3.7kg per growing cycle, 37 kg harvest is expected per year for a VG occupying a space of 1 m^2 which may be difficult to attain if pechay would be grown conventionally (i.e., on the ground). At a prevailing price of Php 40.00/kg, a net income of Php 637.00 per VG could be realized (Table 4).

Thus, the system can potentially provide a good alternative income especially if the crops are grown organically and the harvest be processed for added value like ‘kimchi’ (Korean food) or other kind of processed vetables.

Table 4. Cost and return analysis

	Annual cost, Php
A. Gross income	1, 480.00
B. Fixed cost	
1. Depreciation	243.00
C. Variable cost	
1. Wick replacement every two growing cycle	200.00
2. Seeds	400.00
Sub-total	600.00
D. Total cost	843.00
E. Net income	637.00
F. Gross margin	880.00
G. Payback period (years)	3.6

CONCLUSION

With climate change threatening the sustainability of the Philippine agriculture, there is a need to develop technologies that could help farmers cope up with extreme events like droughts and floods, among others. Nowadays, farmers need to diversify their sources of income and not just rely from the income they get from rice production which is highly climate sensitive. For years now, *Palayamanan*, a diversified and integrated system of rice-based farming is being promoted as a climate change adaptation strategy wherein farmers are encouraged to utilize and maximize the use of vacant spaces in their farms for various income-generating activities like growing of short duration crops such as vegetables and other cash crops especially during extreme weather events (floods and droughts). It is for these reasons that the *capillarigation* system, a do-it-yourself type irrigation system that makes use of capillary wicks as means of dispensing water (same function of the emitter in the drip irrigation system) was developed. This system had been tried and proven to work under drought condition through series of trials done since 2015. In this study, its application in vertical garden was explored. Using the VG model developed in this study and from the results of the test runs conducted, it can be concluded that the *capillarigation* system can be integrated and worked satisfactorily as a system for applying water in vertical gardens. With its integration, the VG can be made self-watering, like in the case of the VG model developed in this study which, with an occupied space of $1m^2$, can accommodate 320 plants (pechay) and able to produce a marketable yield of 3.7 kg in a 30-day growing cycle. Follow up studies is recommended to be done to determine the appropriate nutrient management practice suited for this kind of vegetable production.

Acknowledgments

The authors wish to acknowledge the Philippine Rice Research Institute for funding this research study.

References

[1] AbumhadI, N., Todorovska, E., Assenov, B., Tsonev, S., Vulcheva, D., Vulchev, D., Atanasova, L., Savova, S., and Atanassov, A. 2012. Agricultural Research in 21st century: Challenges facing the food security under the impacts of climate change. *Bulg. J. Agric. Sci.,* 18: 801-818.

[2] Brklacich, M., McNabb, D., Bryant, C. and Dumanski, J. 1997. Adaptability of agriculture systems to global climate change: A

Renfrew County, Ontario, Canada pilot study. In Ilbery, B., Chiotti, Q and Rickard, T. (eds.), *Agricultural Restructuring and Sustainability: A geographical perspective*. Wallingford, CAB International, 351–364.

[3] Corales, R. G., Juliano, L. M., Capistrano, A. O. V., Tobias, H. S., Dasalla, N. V., Cañete, S. D., Casimero, M. C. & Sebastian, L.S. 2004. Palayamanan: a rice-based farming systems model for small-scale farmers. *Philippine Journal of Crop Science,* 29(1), 21-27.

[4] de Loë, R., Kreutzwiser, R. and Mararu, L. 1999. Climate change and the Canadian water sector: Impacts and adaptation, Guelph, Natural Resources Canada.

[5] Food and Agriculture Organization of the United Nations. 2017. *El Nino and La Nina in the Philippines*. Retrieved 4 March 2018 from http://www.fao.org/3/a-i6775e.pdf.

[6] Kelly, P. M. and Adger, W. N. 2000. Theory and practice in assessing vulnerability to climate change and facilitating adaptation. *Climatic Change.* 47(4), 325-352.

[7] Orge, R.F. and Sawey, D.A. 2017. Coping with water scarcity in rice-based farms: development of a capillary irrigation system for small holder farmers in the Philippines. In Gorawala, P. and Mandhatri, S. (eds). *Agricultural Research Updates,* 16, Nova Science Publishers, New York, p. 191-207.

[8] Orge R. F. and Sawey, D. A. 2019. Field performance of the capillary wick irrigation (*capillarigation*) system for rice-based crops. *International Journal of GEOMATE* (in press) ISSN: 2186-2982 (Print), 2186-2990 (Online), Japan.

[9] Smit, B. and Skinner, M. W. 2002. Adaptation options in agriculture to climate change: a typology. *Mitigation and Adaptation Strategies for Global Change.* 7(1), 85-114.

[10] Smithers, J. and Smit, B. 1997. Agricultural system response to environmental stress. In Ilbery, B., Chiotti, Q. and Rickard, T. (eds.), *Agricultural restructuring and sustainability: A geographical perspective*, Wallingford, CAB International, pp. 167–183.

[11] Tuong T. P., Bouman B. A. M. 2003. Rice production in water-scarce environments. In: *Proc. Water Productivity Workshop*, 12-14 November 2001, Colombo, Sri Lanka. International Water Management Institute, Colombo, Sri Lanka.

[12] Vikas, M. and Dwarakish, G. S. 2015. El Niño: A Review. *International Journal of Earth Sciences and Engineering,* pp 130 – 137.

In: Agricultural Research Updates. Vol. 28 ISBN: 978-1-53616-253-0
Editors: P. Gorawala, S. Mandhatri

Chapter 2

Function of Sugar Metabolism, Cell Wall–Loosening Proteins, and Aquaporins in Cut Rose Flowering

Takanori Horibe*
College of Bioscience and Biotechnology, Chubu University,
Kasugai, Aichi, Japan

Abstract

The rose (genus *Rosa*) is one of the msost important plants for ornamental cut flowers and use in ornamental gardens; therefore, understanding the mechanism by which rose flowers open would benefit both general plant science and commercial horticulture. Although many studies have focused on senescence to improve the vase life of cut flowers, there are few studies on the physiology of flower opening. The vase life of cut roses depends, at least in part, on the timing of flower opening. To increase the ornamental value of cut roses, it is desirable that

* Corresponding Author's E-mail: t-horibe@isc.chubu.ac.jp.

the timeframe for complete flower opening be relatively slow; therefore, understanding the mechanism by which rose flowers open is necessary to improve flower quality. In this review, we describe recent progress and findings on the physiology of petal growth in cut rose flowers.

Keywords: aquaporin, cell wall loosening, sugar metabolism, *Rosa*, postharvest

1. Introduction

The rose (genus *Rosa*) is one of the most important ornamental plants for beautiful cut flowers and holds an important position as use in ornamental landscaping. The overall use of roses has a tremendous impact on this industry. Humans are said to have been cultivating roses since before the Common Era, making it one of the most familiar and cherished flowers known to humankind. Initially used in medicine and as a fragrance, the rose eventually came to be appreciated as an ornamental flower and has since evolved as a horticultural product. In the Northern Hemisphere alone, there are approximately 150 species of roses growing in the wild. Although there are tens of thousands of cultivars, only 8 to 20 are considered to have been involved in the breeding of the four main genealogical lines currently being cultivated (Bendahmane et al., 2013). These four main strains of roses are the hybrid tea, which has a single large bloom; floribunda, which has a cluster of medium-sized blooms; climbing, which has tendrils; and miniature, which has small dwarf blooms. Cut flower cultivars are becoming more diverse and range from the more traditional varieties with tall, pointy petals to those with either rounded, cup-shaped petals, old rose–like quarter-bloom petals, or single-petal flowers.

The quality of a cut flower is contingent on the external characteristics of the plant, such as color, length, volume, freshness, and fragrance, as well as perishability, which is determined by the duration of these characteristics. In particular, vase life is a critical factor in determining the market value of cut flowers. Unfortunately, it takes several days before cut flowers produced by farmers reach consumers through retailers;

consequently, the flowers continue to age during the process. As a result, within ~1 week, most cut flowers are no longer suitable for ornamental displays, making it difficult to enjoy these products for an extended period of time. Research has indicated that the factors in shortening the life expectancy of cut flowers include the aging. These include detachment and withering of tissues by ethylene generation, obstruction of vessels due to bacterial propagation (poor water absorption), and deficiency in saccharide (sugar) that act as osmoregulator and respiratory substrate (Durkin 1979; Burdett 1970; De Stigter 1980; Macnish 2010; Reid 1989; van Doorn 1997). Ethylene, in particular, is a huge factor in the aging of many cut flowers; consequently, several studies are being conducted to either suppress ethylene generated during the blooming of cut flowers or inhibit its action to slow the aging process (Ichimura et al., 2007).

As a result of continued efforts to improve the quality of cut flowers and ensure that they last a long time, many agents to preserve that quality have been developed, such as ethylene inhibitors. It has been reported that ethylene in the rose flower regulates its opening and petal expansion as well as its senescence (Ma et al. 2006; Ma et al. 2008; Tan et al. 2006). Although Van Doorn and Kamdee (2014) have reviewed the relationship between flower opening and hormones, many other studies have concentrated only on ways by which to decelerate the aging process (Ichimura and Shimizu-Yumoto 2007). For ornamental plants, especially flowers such as roses, whose ornamental value lies in the process of blooming from its buds, it is more important to understand their flowering mechanisms and establish methods by which to control those mechanisms than to focus on aging.

2. Flower Opening and Petal Cell Morphology

The process of the rosebud development and subsequent flower opening involves irreversible petal growth and reflection in which existing cells expand and fresh and dry weights increase (Reid and Evans, 1986; Evans and Reid, 1988; Faragher et al., 1984). In carnation flowers, the amount of DNA in the petals does not increase once the petals emerge from the calyx, which suggests that cells also stop dividing at an early stage (Kenis et al. 1985). In *Gaillaradia grandiflora*, petal cells appear to stop dividing at a much earlier flowering stage, and no increase in the number of abaxial epidermal cells has been observed (Koning, 1984). These findings indicate that petal growth related to flower opening depends mainly on cell expansion. Cells in rose petals almost completely stop dividing even when the petals remain covered by the calyx (Roberts et al. 2003; Yamada et al., 2009c), which suggests that petal growth associated with flower opening depends on cell expansion. Sugar accumulation in the vacuole, cell wall loosening, and the subsequent water flow into the cell is believed to be important factors in the enlargement of rose petals (Figure 1) (Horibe and Yamada, 2017).

The adaxial and abaxial epidermal cells in the cut rose show marked expansion during flower opening, and adaxial epidermal cells become corn shaped when the flower is fully open (Yamada et al., 2009c). Differences in the patterns of cell expansion among cell types and locations, including adaxial and abaxial epidermal cells, are believed to cause petal reflection during rose flower opening. Cortical microtubules have been proposed to regulate the orientation of cell expansion (Shibaoka et al., 1994); however, ethylene inhibits longitudinal cell expansion and promotes lateral expansion, which is associated with changes in cortical microtubule orientation (Roberts et al., 1985; Steen and Chadwick, 1981). In some rose cultivars, ethylene accelerates flower opening (Reid et al., 1989; Tan et al., 2006) and treatment, while the silver thiosulfate complex suppresses petal reflection (Doi, 1995). Thus, the lateral growth of adaxial epidermal cells might be regulated through endogenous ethylene production by changing the orientation of the cortical microtubules. Although exposure to high concentrations of ethylene induces petal abscission in *Rosa* “Sonia” (Ichimura et al., 2005; Reid et al., 1989), endogenous ethylene production in this cultivar is relatively low (Ichimura et al., 2005), which does not rule out the possible involvement of ethylene in lateral cell expansion in the petals.

Under an optical microscope and using scanning electron microscopy, the parenchyma cells of rose petals showed unique growth into a tetrapod-like shape that resembles the mesophyll cells of leaves (Yamada et al., 2009c). Mesophyll cell morphogenesis and intercellular space formation were investigated in the leaves of *Adiantum capillus-veneris* (Panteris et al., 1993) and *Nigella damascene* (Wernicke et al., 1993), and the results revealed that mesophyll cell morphogenesis and intercellular space formation are accurately controlled by highly organized cortical microtubule systems, with the disposition of their bundles among neighboring mesophyll cells being highly coordinated (Hellmann and Wernicke, 1998; Schroder et al., 2001; Uehara and Hogetsu, 1993). Intercellular space formation in leaves appears to allow space for gas exchange during photosynthesis. Although petal tissues are similar to those in leaves, their functions are quite different; therefore, the role of the air space in rose petals with regard to flower opening is unclear.

3. Sugars and Cut Flower Opening

Large amounts of soluble sugars accumulate in rose petals, especially in the vacuoles within the rose petal cells during flower opening (van Doorn et al. 1991; Ichimura et al. 2003). Several studies have reported that sugars are a factor in extending the vase life of cut flowers (Paulin 1979; Paulin and Jamain 1982; Ichimura et al. 2003), and that sucrose treatment increases the vase life of cut carnations and roses (Paulin and Jamain, 1982; Ichimura et al., 2003), which suggests that soluble carbohydrates play an important role in regulating osmotic pressure in petal cells (Figure 1). It is reasonable to presume that sugar accumulation in petal cells leads to a reduction in water potential and promotes water influx, which might lead to cell enlargement and flower opening (Ho and Nichols 1977). In tulip petals, glucose and fructose accumulate mainly in the vacuoles (Wagner 1979). In addition to accumulating in the vacuoles, soluble carbohydrates accumulate in the apoplast in some sugar sink organs (e.g., roots, tubers), such as tomato fruits (Damon et al. 1988) and sugarcane stalks (Welbaum and Meinzer 1990). Flowers are also sink organs, and the plant's carbohydrates accumulate in the apoplasts of rose petal cells (Yamada et al., 2009b). To accelerate water influx into the cells, osmotic pressure should be higher in the symplast than in the apoplast.

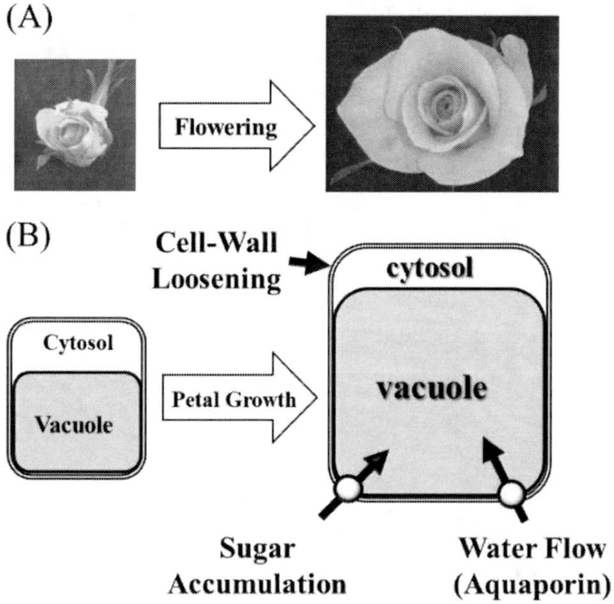

Figure 1. Flower opening and enlargement of rose petal cells. Rose flower opening is a process of irreversible petal growth and reflection in which existing cells expand and fresh and dry weights increase (A). Sugar accumulation in the vacuole, cell wall loosening, and subsequent water flow into the cell are believed to be important for cell enlargement (B).

In most rose species, starch hydrolysis is important for promoting rose petal growth (Hammond, 1982), and it has been suggested that the degradation of stored carbohydrates, the import of sucrose, or both are associated with flower opening (van Doorn and van Meeteren, 2003). Young petal cells of many species contain considerable amounts of starch, which is rapidly converted into hexose (Hammond, 1982; Ho and Nichols, 1977). Petal starch levels at harvest are high in many rose cultivars but low in others, such as "Madelon" (Berkholst 1989; Gorin and Berkholst 1982; van Doorn et al., 1991; van Doorn and van Meeteren 2003). Gorin and

Berkholst (1982) have elaborated that a criterion of rose maturity for harvest is the starch content in petals as follows: <10% on a dry-weight basis would represent that flowers were harvested too early. Kuiper et al. (1995) have reported that adding sucrose to cut "Madelon" buds induced proper flower opening, even though the cut flowers of this variety frequently fail to open completely under postharvest conditions, which indicates the importance of sugar accumulation in the petal for petal growth. However, it has been reported that transported sucrose appears to contribute more to reducing the starch reserved in the rose petals before flower opening because the increased amount of soluble sugars in the petals during this time was much higher than the amount of starch (Ichimura et al. 1999; Yamada et al., 2009a). Thus, the increase in monosaccharide content derived from the photosynthetic product transported from the leaves appears to be important for proper flower opening in some rose cultivars. In some other flowers, such as the daylily (*Hemerocallis*) (Bieleski 1993) and *Campanula rapunculoides* (Vergauwen et al. 2000), fructans, which are oligosaccharides and polysaccharides, in the petals have been shown to decrease during flower opening, which is associated with an increase in monosaccharides.

In addition to glucose, fructose, and sucrose, which are ubiquitous metabolic sugars, methyl glucoside, xylose, and myo-inositol are soluble carbohydrate constituents in roses (Ichimura et al. 1997). Myo-inositol increases osmotic pressure and acts as a precursor to cell wall synthesis (Loewus and Dickinson 1982), but its application inhibits flower opening in cut roses (Ichimura et al. 1999). The applied myo-inositol might accumulate in the apoplast and suppress water influx into the vacuole, which would inhibit cell expansion associated with flower opening. Although the application of methyl glucoside and xylose promotes flower opening in cut roses (Ichimura et al. 1999), their concentrations were relatively low and did not increase during rose flower opening (Yamada et al., 2009a). Thus, these carbohydrates appear to not be important for flower opening in the rose plant.

In the shoot and leaves of some plants, K^+ is a main ion that contributes to osmotic pressure; however, Ca^{2+}, Mg^{2+}, NO_3^-, and Cl^- are also major ions (Cram 1976; Leigh and Tomos 1993). In addition, organic acids have been shown to contribute to the increase in osmotic pressure in many plants, such as sugarcane stalk (Cram 1976; Welbaum and Meinzer 1990). Malic acid and citric acid are the major organic acids in rose petals (Biran et al. 1974; Schnabl and Mayer 1976). Yamada et al. (2009a) have determined the subcellular distribution of soluble carbohydrates and inorganic ions by combining the nonaqueous fractionation method with the infiltration-centrifugation method, and have shown that the increase in osmotic pressure in apoplasts and symplasts during flower opening is mainly attributed to increases in fructose and glucose concentrations.

Vase life is one of the most important factors that determine the quality of cut flowers. The vase life varies among various cultivars of the carnation (Wu et al., 1991), gerbera (Wernett et al., 1996), and rose (Ichimura et al., 2002). The vase life of cut rose flowers often terminates when the floral axis is bent just below the flower head, called "bent neck." Parups and Voisey (1976) have shown that pedicels thicken at a relatively late stage in flower development and an insufficiently lignified pedicel leads to bent neck. An insufficiently lignified pedicel remains rigid only because of cell turgidity and the smallest turbulences in water balance are manifested within this region of a cut rose stem, producing bent neck symptoms. They have also reported that spraying rose shrubs with phenolic compounds before harvest could promote lignification and prevent this phenomenon (Parups and Voisey 1976). Vascular occlusion is also thought to cause the development of bent neck, which restricts the water supply to the flowers (Van Doorn 1997). Vascular occlusion is caused by increases in bacteria (Van Doorn et al. 1989; Jones and Hill 1993), air emboli (Durkin 1979), or unknown physiological responses (Marousky 1969). Van Doorn and D'hont (1994) have reported that resistance to bacteria and dry storage are involved in variations in the vase life of cut rose cultivars.

Rose flowers are harvested during the bud stage, and their vase life, an important characteristic of flower quality, is relatively short. Cut roses held in water lack soluble carbohydrates, which results in the suppression of petal growth and incomplete flower opening (Ichimura et al., 2003). However, adding sugars to the vase solution improves the quality of various cut flowers, including the rose (van Doorn et al. 1991; Ichimura et al. 2003). Furthermore, adding sucrose to the solution increases the levels of glucose and fructose, but has little effect on sucrose content in the petals, which indicates that sucrose that translocates into the petals from other organs is metabolized into glucose and fructose and accumulates in petal cells (Kaltaler and Steponkus, 1976). Soluble carbohydrates are believed to act as osmotica and substrates for both respiration and cell wall synthesis for cell expansion (Evans and Reid, 1988; Ichimura et al. 2003). In the cut rose flower, the leaves play an important role in maintaining water uptake rates through transpiration (Halevy and Mayak 1981), and the removal of these leaves reduces this rate and prevents flowers from fully opening. It has also been shown that the ability of the stomates to close in response to water stress conditions was a decisive factor in postharvest longevity (Mayak, 1974). Adverse water relationships in the rose are associated with incomplete flower opening, premature petal wilting, and bending of the pedicel (Doi et al. 1999). The leaves and stems of cut flowers can also act as a source of soluble carbohydrates for the flower sink tissue. In the chrysanthemum, soluble carbohydrates that accumulate in the leaves and stems were reported to affect the vase life of the cut flower (Ishikawa et al. 2006). In the cut rose, soluble carbohydrates that accumulated in the petals also affected its vase life (Ichimura et al., 2005). Horibe et al. (2014) have shown that most of the exogenously applied glucose first moves to the leaves where it is converted into sucrose. Then, the synthesized sucrose translocates to the petals, as reported in *Eustoma* (Shimizu-Yumoto and Ichimura 2007). Chin and Sacalis (1977) have reported that 14C-sucrose, when taken up into the xylem, moves rapidly into the leaves and flower heads of cut roses, and that the invertase located in the xylem is involved in the rapid hydrolysis of sucrose. Paulin and Jamain (1982) have also investigated the interconvention of sugars during transport in the rose and carnation. An abundance of soluble carbohydrates accumulate in the leaves after sugar treatment, but only a portion of these appears to translocate into the petals (Horibe et al., 2014); therefore, promoting sugar translocation from the leaves to the petals would improve the quality of cut flowers and elevate the positive effects of sugar treatment. More studies are needed to further understand the functions of the leaves of cut flowers with regard to sugar metabolism and sucrose translocation.

4. Function of Sugar-Metabolizing Protein

There are two pathways for phloem unloading—apoplastic and symplastic (Patrick and Offler 1996). After unloading, the sugars are transported through the apoplast and/or symplast pathway. Sucrose synthase and acid invertase (β-fructofuranosidase), which are present in the vacuole (soluble form) and cell wall (insoluble form) in many higher plants, are the enzymes that metabolize the sucrose that is translocated from the leaves into the sink tissues. The acid invertase in the cell wall plays a role in converting sucrose into hexoses after sucrose is translocated from the phloem to the apoplast (Roitsch and Gonzalez-Garcia 2004), which enables the petals to uptake more sucrose from the phloem. Vacuolar invertase also plays an important role in the biological functions associated with sucrose metabolism by presumably hydrolyzing sucrose to supply the hexoses necessary for cell growth and development (Tang et al. 1999). It has been shown that invertase activity markedly increases during petal growth in the petals of attached rose flowers, but not in cut roses, even when cut roses are treated with sucrose (Yamada et al. 2007). On the other hand, sucrose synthase activity is very weak compared with that of acid invertase in rose petals (Kumar et al. 2008), which suggests that sugars appear to be transported, at least partially, through apoplastic unloading in rose petals. Thus, high glucose and fructose concentrations in the apoplast might be caused from the transport of these carbohydrates through the apoplast (Yamada et al. 2007). In apple (Zhang et al. 2004) and tomato fruits (Damon et al. 1988), in which carbohydrate concentrations in the apoplasts are relatively high, the sugars are partially transported through the apoplast. Numerous reports have indicated a relationship between invertase activity and the strength of sink tissues (Tang et al. 1999; Balibrea et al. 2004; Roitsch and Gonzalez-Garcia 2004). The invertase activity appears to limit petal growth and is related to flower opening. Some studies have shown that certain phytohormones affect invertase activity (Miyamoto et al. 1993; Trouverie et al. 2004; Pan et al. 2006; González and Cejudo, 2007). The acid invertase activity was also affected by NAA and MeJA treatment in the cut rose and, interestingly, the speed of flower opening also changed with each treatment (Horibe et al., 2013). Thus, the quality of cut rose flowers might be improved by controlling invertase activity under postharvest conditions.

5. Function of Cell Wall–Loosening Proteins

Cell expansion is regulated not only by the accumulation of sugars and water influx, but also by cell wall pressure, which is determined by wall strength. When the cell wall of petals remains rigid, water cannot influx into the cell and the cell cannot expand. Thus, cell wall loosening is also important for petal cell enlargement (Figure 1). Cell wall extensibility might be a growth-limiting factor for rose petal expansion. Until recently, cell wall extensibility during flower opening has received little attention; however, cell wall loosening during flower opening is an important characteristic and a precursor to detailed molecular investigation.

Many proteins and enzymes are required for cell expansion. Cell wall–loosening candidate proteins include endotransglycosylase/hydrolase (XTH) and expansins (Cosgrove, 2001). Expansins were first identified in cell wall–protein fractions that restored acid-induced extension of heat-inactivated cell walls (McQueen-Mason et al., 1992). They are believed to disrupt hydrogen bonding between cellulose microfibrils and matrix glucans (McQueen-Mason and Cosgrove, 1994). Expansin genes belong to a multigene superfamily that includes two families of genes—α-expansin, which was identified first, and β-expansin (Cosgrove, 2001)—and are commonly identified in expanding and growing tissues. Expansins are expressed in nongrowing tissue, such as that in ripening fruits (Brummel et al., 1999; Civello et al., 1999; Rose et al., 1997) and the abscission zone (Belfield et al., 2005; Sane et al., 2007). XTHs were first identified and characterized in higher plants (Nishitani and Tominaga 1992). Each gene in the XTH family catalyzes either molecular grafting or the disassembly of xyloglucan crosslinks within the cellulose–xyloglucan framework (Nishitani and Tominaga, 1992; Okazawa et al., 1993); therefore, XTHs participate in loosening and rearranging the cell walls in growing tissues (Rose et al., 2002) and might also participate in cell expansion (Nishitani, 1997; Vissenberg et al., 2000). XTHs are expressed in growing tissues and ripening fruits (Catalá et al., 2000; Rose and Bennett, 1999) as well as in developmental processes, such as aerenchyma formation (Antosiewicz et al., 1997; Saab and Sachs, 1996).

In ornamental plants, α- and β-expansin were identified from the opening and senescing flowers of *Mirabilis jalapa* (Gookin et al., 2003). Two expansin genes are also associated with petal growth and development during carnation flower opening (Harada et al., 2011). In addition, expansin is expressed during flower opening of tomato (Brummel et al., 1999) and *Petunia hybrida* (Zenoni et al., 2004), and XTH activity is also associated with the opening of sandersonia flowers (O'Donoghue et al., 2002). In roses, *RhEXPA*, one of three α-expansin genes (*RhEXPA1–RhEXPA4*), and *RhXTH1*, one of four endotransglycosylase/hydrolase genes (*RhXTH1–RhXTH4*) are mainly involved in petal expansion (Takahashi et al., 2007; Yamada et al., 2009b; Dai et al., 2012). It has been reported that *RhEXPA4* is involved in the regulation of dehydration tolerance during rose petal expansion (Dai et al., 2012). Sane et al. (2007) have cloned *RbEXPA1* from *R. bourboniana* and have shown that *RbEXPA1* is expressed in senesced petals, which suggests its correlation with petal abscission.

An increase in cell wall extensibility by the action of expansin and XTH, coupled with an increase in osmotic pressure through invertase activity, might facilitate the rapid expansion of rose petal cells. However, additional research is necessary to better understand the relationship between cell wall–loosening proteins and petal growth.

6. FUNCTION OF AQUAPORINS

Aquaporins are multifunctional membrane channels that facilitate the transportation of water molecules and low molecular weight compounds across membranes (Katsuhara et al., 2008). Based on similarities in amino acid sequences, aquaporins are classified into the following seven subfamilies: plasma membrane intrinsic proteins (PIPs) and tonoplast intrinsic proteins (TIPs), which are the most abundant aquaporins in the plasma membrane and tonoplast; nodulin 26-like intrinsic proteins (NIPs), which are located in the peribacteroid membrane of nitrogen-fixing symbiotic root nodules of leguminous plants and in the plasma membrane of other species; small basic intrinsic proteins, which are small proteins mainly localized in the endoplasmic reticulum membrane; uncharacterized X intrinsic proteins, which are plasma membrane aquaporins that function in the transport of uncharged substrates; and hybrid intrinsic proteins and glycerol facilitator (GlpF)-like intrinsic proteins (GIPs), which are present exclusively in moss (Wang et al., 2016). Among these aquaporins, PIPs and TIPs are believed to play key roles in plant growth and cell enlargement (Katsuhara et al., 2008).

Azad et al. (2004; 2008; 2013) have suggested that petal opening and closure in the tulip occur concomitantly with water transport and are regulated by reversible phosphorylation of PIPs, and that two TIPS (*TgTIP1;1* and *TgTIP1;2*) contribute to petal development. In the rose, it has been shown that the expression of *Rh-TIP1;1* is related to stress reactions to ethylene and water deficits, and that *RhPIP1;1* silencing significantly inhibits petal expansion together with decreased petal size and cell area, thus demonstrating the important role of aquaporins in petal cell expansion (Chen et al., 2013). Because aquaporins function in the transport of water and low molecular weight compounds, it is plausible that they also play an important role in flower opening; however, relatively little is known about their functions in rose flower opening.

Conclusion

In this review, we have described cell morphology, sugar metabolism, cell wall–loosening proteins, and aquaporins as they relate to cut rose flower opening; however, the specific interrelationships among the subject areas remain unclear. Research has been conducted that focused on postharvest physiology, including flower senescence, which has led to the invention of ethylene production/action inhibitors; however, few studies have focused on the physiology of petal growth. Determining and clarifying the mechanisms by which flowers open would contribute to the commercial production of floricultural crops as well as to general plant science.

References

Antosiewicz, D. M., Purugganan, M. M., Polisensky, D. H. & Braam, J. (1997). Cellular localization of Arabidopsis xyloglucan endotransglycosylase-related proteins during development and after wind stimulation. *Plant Physiol.*, *115*, 1319-1328.

Azad, A. K., Sawa, Y, Ishikawa, T. & Shibata, H. (2004). Phosphorylation of plasma membrane aquaporin regulates temperature-dependent opening of tulip petals. *Plant Cell Physiol.*, *45*, 608-617.

Azad, A. K., Katsuhara, M., Sawa, Y., Ishikawa, T. & Shibata, H. (2008). Characterization of four plasma membrane aquaporins in tulip petals: a putative homolog is regulated by phosphorylation. *Plant Cell Physiol.*, *49*, 1196-1208.

Azad, A. K., Hanawa, R., Ishikawa, T., Sawa, Y. & Shibata, H. (2013). Expression profiles of aquaporin homologues and petal movement during petal development in Tulipa gesneriana. *Physiol Plant.*, *148*, 397-407.

Balibrea, L., Gonzalez-Garcia, M. E., Fatima, M. C., Ehneb, T., Lee, R., Proels, T. K., Tanner, R. W. & Roitsch, T. (2004). Extracellular invertase is an essential component of cytokinin-mediated delay of senescence. *Plant Cell.*, *16*, 1276-1287.

Belfield, E. J., Ruperti, B., Roberts, J. A. & McQueen-Mason, S. J. (2005). Changes in expansin activity and gene expression during ethylene-promoted leaflet abscission in Sambucus nigra. *J. Exp. Bot.*, *56*, 817-823.

Bendahmane, M., Dubois, A., Raymond, O. & Bris, M. L. (2013). Genetics and genomics of flower initiation and development in roses. *J. Exp. Bot.*, *64*, 847-857.

Berkholst, C. E. M. (1989). High starch content in 'Sonia' rose corollas at picking may add quality to vase life. *Gartenbauwissenschaft*, *54*, 9-10.

Bieleski, R. L. (1993). Fructan hydrolysis drives petal expansion in the ephemeral daylily flower. *Plant Physiol.*, *103*, 213-219.

Biran, M., Robinson, M. & Halevy, A. H. (1974). Factors determining petal color of Baccara roses. *J. Exp. Bot.*, *87*, 624-663.

Brummel, D. A., Harpster, M. H. & Dunsmuir, P. (1999). Differential expression of expansin gene family members during growth and ripening of tomato fruit. *Plant Mol. Biol.*, *39*, 161-169.

Burdett, A. N. (1970). The cause of bent neck in cut roses. *J. Amer. Soc. Hort. Sci.*, *95*, 427-431.

Catalá, C., Rose, J. K. C. & Bennett, A. B. (2000). Auxin-regulated genes encoding cell wall-modifying proteins are expressed during early tomato fruit growth. *Plant Physiol.*, *122*, 527-534.

Chen, W., Yin, X., Wang, L., Tian, J., Yang, R., Liu, D., Yu, Z., Ma, N. & Gao, J. (2013). Involvement of rose aquaporin RhPIP1; 1 in ethylene-regulated petal expansion through interaction with RhPIP2;1. *Plant Mol Biol.*, *83*, 219-33.

Chin, C. & Sacalis, J. N. (1977). Metabolism of sucrose in cut roses. II. Movement and inversion of sucrose absorbed by cyt rose stems. *J. Am. Soc. Hort. Sci.*, *102*, 537-540.

Civello, P. M., Powell, A. L. T., Sabehat, A. & Bennett, A. B. (1999). An expansin gene expressed in ripening strawberry fruit. *Plant Physiol.*, *121*, 1273-1279.

Cosgrove, D. J. (2001). Wall structure and wall loosening. A look backwards and forwards. *Plant Physiol.*, *125*, 131-134.

Cram, W. J. (1976). Negative feedback regulation of transport in cells. The maintenance of turgor, volume and nutrient supply. In: Lu¨ttge U, Pitman MG (eds) *Encyclopedia of plant physiology*, N. S., vol *2*, transport in plants II, part A. Springer, Berlin, pp. 284-316.

Dai, F., Zhang, C., Jiang, X., Kang, M., Yin, X., Lü, P., Zhang, X., Zheng, Y. & Gao, J. (2012). RhNAC2 and RhEXPA4 are involved in the regulation of dehydration tolerance during the expansion of rose petals. *Plant Physiol.*, *160*, 2064-2082.

Damon, S., Hewitt, J., Nieder, M. & Bennett, A. (1988). Sink metabolism in tomato fruit. II. Phloem unloading and sugar uptake. *Plant Physiol.*, *87*, 731-736.

De Stigter, H. C. M. (1980). Water balance of cut and intact Sonia rose plants. *Z. Pflanzenphysiol.*, *99*, 131140.

Doi, M. (1995). Flower durability. In: *Nogyo Gijutsu Taikei, Kakihen*, vol. *3*. Nobunkyo, Tokyo, pp. 105-109.

Doi, M., Miyagawa Namao, M., Inamoto, K. & Imanishi, H. (1999). Rhythmic changes in water uptake, transpiration and water potential of cut roses as affected by photoperiods. *J. Japan. Soc. Hort. Sci.*, *68*, 861-867.

Durkin, D. J. (1979). Effect of Millipore filtration, citric acid, and sucrose on peduncle water potential of cut rose flower. *J. Amer. Soc. Hort. Sci.*, *104*, 860-863.

Evans, R. Y. & Reid, M. S. (1988). Changes in carbohydrates and osmotic potential during rhythmic expansion of rose petals. *J. Amer. Soc. Hort. Sci.*, *113*, 884-888.

Faragher, J. D., Mayak, S., Tirosh, T. & Halevy, A. H. (1984). Cold storage of rose flowers: Effects of cold storage and water loss on opening and vase life of 'Mercedes' roses. *Sci. Hort.*, *24*, 369-378.

Gonzalez-Garcia, M. C. & Cejudo, F. J. (2007). Gibberellin-regulated expression of neutral and vacuolar invertase genes in petioles of sugar beet plants. *Plant Sci.*, *172*, 839-846.

Gookin, T. E., Hunter, D. A. & Reid, M. S. (2003). Temporal analysis of alpha- and beta-expansin expression during floral opening and senescence. *Plant Sci.*, *164*, 769-781.

Gorin, N. & Berkholst, C. E. M. (1982). Starch in petals of cut roses cv. Sonia as possible criterion of picking. *Gartembauwissenschaft*, *47*, 75-77.

Halevy, A. H. & Mayak, S. (1981). Senescence and postharvest physiology of cut flowers: Part 2. *Hort. Reviews.*, *3*, 59-143.

Hammond, J. B. W. (1982). Changes in amylase activity during rose bud opening. *Sci. Hort.*, *16*, 283-289.

Harada, T., Torii, Y., Morita, S., Onodera, R., Hara, Y., Yokoyama, R., Nishitani, K. & Satoh, S. (2011). Cloning, characterization, and expression of xyloglucan endotransglucosylase/hydrolase and expansin genes associated with petal growth and development during carnation flower opening. *J. Exp. Bot.*, *62*, 815-823.

Hellmann, A. & Wernicke, W. (1998). Changes in tubulin protein expression accompany reorganization of microtubular arrays during cell shaping in barley leaves. *Planta*, *204*, 220-225.

Ho, L. C. & Nichols, R. (1977). Translocation of ^{14}C-sucrose in relation to changes in carbohydrate content in rose corollas cut at different stages of development. *Ann. Bot.*, *41*, 227-242.

Horibe, T., Yamaki, S. & Yamada, K. (2013). Effects of auxin and methyl jasmonate on cut rose petal growth through activation of acid invertase. *PBT*, *86*, 195-200.

Horibe, T., Yamaki, S. & Yamada, K. (2014). Leaves of cut rose flower convert exogenously applied glucose to sucrose and translocate it to petal. *J. hort. Research*, *22*, 41-46.

Horibe, T. & Yamada, K. (2017). Petal growth physiology of cut rose flowers: progress and future prospects. *J. hort. Research*, *25*, 5-18.

Ichimura, K., Kohata, K., Koketsu, M., Yamaguchi, Y., Yamaguchi, H. & Suto, K. (1997). Identification of methyl glucopyranoside and xylose as soluble sugar constituents in roses (Rosa hybrida L.). *Biosci Biotechnol Biochem*, *61*, 1734-1735.

Ichimura, K., Mukasa, Y., Fujiwara, T., Kohata, K., Goto, R. & Suto, K. (1999). Possible roles of methyl glucoside and myoinositol in the opening of cut rose flowers. *Ann. Bot.*, *83*, 551-557.

Ichimura, K., Kawabata, Y., Kishimoto, M., Goto, R. & Yamada, K. (2002). Variation with the cultivar in the vase life of cut rose flowers. *Bull. Natl. Inst. Flor. Sci.*, 2, 9-20.

Ichimura, K., Kawabata, Y., Kishimoto, M., Goto, R. & Yamada, K. (2003). Shortage of soluble carbohydrates is largely responsible for short vase life of cut 'Sonia' rose flowers. *J. Japan. Soc. Hort. Sci.*, *72*, 292-298.

Ichimura, K., Kishimoto, M., Norikoshi, R., Kawabata, Y. & Yamada, K. (2005). Soluble carbohydrates and variation in vaselife of cut rose cultivars 'Delilah' and 'Sonia'. *J. Hort. Sci. Biotechnol.*, *80*, 280-286.

Ichimura, K. & Shimizu-Yumoto, H. (2007). Extension of the vase life of cut roses by treatment with sucrose before and during simulated transport. *Bull. Natl. Inst. Flor. Sci.*, *7*, 17-27.

Ishikawa, T., Nishio, J. & Ichimura. K. (2006). Effects of light shielding treatment during cultivation on vase life and sugar content of cut chrysanthemum. *Res. Bull. Aichi Agric. Res. Ctr.*, *38*, 127-132.

Jones, R. B. & Hill, M. (1993). The effect of germicides on the longevity of cut flowers. *J. Amer. Soc. Hort. Sci.*, *118*, 350-354.

Kaltaler, R. E. L. & Steponkus, P. L. (1976). Factors affecting respiration in cut roses. *J. Amer. Soc. Hort. Sci.*, *101*, 352-354.

Katsuhara, M., Hanba, Y. T., Shiratake, K. & Maeshima, M. (2008). Expanding roles of plant aquaporins in plasma membranes and cell organelles. *Funct. Plant Biol.*, *35*, 1-14.

Kenis, J. D., Silvents, S. T. & Trippi, V. S. (1985). Nitrogen metabolite and senescence-associated change during growth of carnation flowers. *Physiol. Plant.*, *65*, 455-459.

Koning, R. E. (1984). The role of plant hormones in the growth of the corolla of Gaillardia grandiflora (Asteraceae) ray flowers. *Am. J. Bot.*, *71*, 1-8.

Kuiper, D., Ribot, S., Van Reenen, H. S. & Marissen, N. (1995). The effect of sucrose on the flower bud opening of 'Madelon' cut roses. *Sci. Hortic.*, *60*, 325-336.

Kumar, N., Srivastava, G. C. & Dixit, K. (2008). Role of sucrose synthase and invertase during petal senescence in rose (Rosa hybrid L.). *J. Hort. Sci. Biotechnol.*, *83*, 520-524.

Leigh, R. A. & Tomos, A. D. (1993). Ion distribution in cereal leaves: pathways and mechanisms. *Phil. Trans. R. Soc. Lond. B.*, *341*, 75-86.

Loewus, F. A. & Dickinson, D. B. (1982). Cyclitols. In: Loewus FA, Tanner W (eds) *Encyclopedia of plant physiology*, N. S., vol 13A: plant carbohydrates I. Springer, Berlin, pp. 193-216.

Ma, N., Tan, H., Liu, X., Xue, J., Li, Y. & Gao, J. (2006). Transcriptional regulation of ethylene receptor and CTR genes involved in ethylene-induced flower opening in cut rose (Rosa hybrida) cv. Samantha. *J. Exp. Bot.*, *57*, 2763-73.

Ma, N., Xue, J., Li, Y., Liu, X., Dai, F., Jia, W., Luo, Y. & Gao, J. (2008). Rh-PIP2;1, a rose aquaporin gene, is involved in ethylene-regulated petal expansion. *Plant Physiol.*, *148*, 894-907.

Macnish A. J., Leonard, R. T., Borda, A. M. & Nell, T. A. (2010). Genetic variation in the postharvest performance and ethylene sensitivity of cut rose flowers. *Hort. Sci.*, *45*, 790-796.

McQueen-Mason, S. J., Durachko, D. M. & Cosgrove, D. J. (1992). Two endogenous proteins that induce cell-wall extension in plants. *Plant Cell*, *4*, 1425-1433.

McQueen-Mason, S. J. & Cosgrove, D. J. (1994). Disruption of hydrogen-bonding between plant-cell wall polymers by proteins that induce wall extension. *Proc. Natl. Acad. Sci.* USA, *91*, 6574-6578.

Marousky, F. J. (1969). Vascular blockage, water absorption, stomatal opening, and respiration of cut 'Better times' roses treated with 8-hydroxiquinoline citrate and sucrose. *J. Amer. Soc. Hort. Sci.*, *94*, 223-226.

Mayak, S., Halevy, A. H., Sagie, S., Bar-Yoseph, A. & Bravdo, B. (1974). The water balance of cut rose flowers. *Physiol. Plant*, *31*, 15-22.

Miyamoto, K., Ueda, J. & Kamisaka, S. (1993). Gibberellin-enhanced sugar accumulation in growing subhooks of etiolated Pisum sativum seedlings. Effects of gibberellic acid, indoleacetic acid and cycloheximide on invertase activity, sugar accumulation and growth. *Physiol. Plant*, *88*, 301-306.

Nishitani, K. & Tominaga, R. (1992). Endo-xyloglucan transferase, a novel class of glycosyltransferase that catalyzes transfer of a segment of xyloglucan molecule to another xyloglucan molecule. *J. Biol. Chem.*, *267*, 21058-21064.

Nishitani, K. (1997). The role of endoxyloglucan transferase in the organization of plant cell walls. *Intl. Rev. Cytology*, *173*, 157- 206.

O'Donoghue, E. M., Somerfield, S. D. & Heyes, J. A. (2002). Organization of cell walls in Sandersonia aurantiaca floral tissue. *J. Exp. Bot.*, *53*, 513-523.

Okazawa, K., Sato, Y., Nakagawa, T., Asada, K., Kato, I., Tomita, E. & Nishitani, K. (1993). Molecular cloning and cDNA sequencing of endoxyloglucan transferase, a novel class of glycosyltransferase that mediates molecular grafting between matrix polysaccharides in plant cell walls. *J. Biol. Chem.*, *268*, 25364-25368.

Pan, Q. H., Yu, X. C., Zhang, N., Zou, X., Peng, C. C., Wang, X. L., Zou, K. Q. & Zhang, D. P. (2006). Activity, but not expression, of soluble and cell wall-bound acid invertase is induced by abscisic acid in developing apple fruit. *J. Integrative Plant Biol.*, *48*, 536-549.

Panteris, E., Apostolakos, P. & Galatis, B. (1993). Microtubule organization, mesophyll cell morphogenesis, and intercellular space formation in Adiantum capillus veneris leaflets. *Protoplasma*, *172*, 97-110.

Patrick, J. W. & Offler, C. E. (1996). Post-sieve element transport of photoassimilates in sink regions. *J. Exp. Bot.*, *47*, 1165-1177.

Parups, E. V. & Voisey, P. W. (1976). Lignin content and resistance to bending of the pedicel in greenhouse grown roses. *J. Hort. Sci.*, *51*, 253-259.

Paulin, A. (1979). Evolucion des glucides dans les divers organs de la rose coupee (var. Carina) alimentee temporairement avec une solution glucose. *Physiologie Veg.*, *17*, 129-143.

Paulin, A. & Jamain, C. (1982). Development of flowers and changes in various sugars during opening of cut carnations. *J. Am. Soc. Hort. Sci., 107*, 258-261.

Reid, M. S. & Evans, R. Y. (1986). Control of cut flower opening. *Acta Hort., 181*, 45-54.

Reid, M. S., Evans, R. Y., Dodge, L. L. & Mor, Y. (1989). Ethylene and silver thiosulfate influence opening of cut rose flowers. *J. Amer. Soc. Hort. Sci., 114*, 436-440.

Roberts, I. N., Lloyd, C. W. & Roberts, K. (1985). Ethylene-induced microtubule reorientations: mediation by helical arrays. *Planta, 164*, 439-447.

Roberts, A. V., Debener, T. & Gudin, S. (2003). Elsevier Academic Press, Amsterdam. *Encyclopedia of rose science 1st edition*, vol 2, pp. 504-512.

Roitsch, T. & Gonzalez-Garcia, M. C. (2004). Function and regulation of plant invertases: sweet sensations. *Trends Plant Sci., 9*, 606-613.

Rose, J. K. C., Lee, H. H. & Bennett, A. B. (1997). Expression of a divergent expansin gene is fruit-specific and ripeningregulated. *Proc. Natl. Acad. Sci. USA, 94*, 5955-5960.

Rose, J. K. C. & Bennett, A. B. (1999). Cooperative disassembly of the cellulose-xyloglucan network of plant cell walls: parallels between cell expansion and fruit ripening. *Trends Plant Sci., 4*, 176-183.

Rose, J. K. C., Braam, J., Fry, S. C. & Nishitani, K. (2002). The XTH family of enzymes involved in xyloglucan endotransglucosylation and endohydrolysis: current perspectives and a new unifying nomenclature. *Plant Cell Physiol., 43*, 1421-1435.

Saab, I. N. & Sachs, M. M. (1996). A flooding-induced xyloglucan endo-transglycosylase homolog in maize is responsive to ethylene and associated with aerenchyma. *Plant Physiol., 112*, 385-391.

Sane, A. P., Tripathi, S. K. & Nath, P. (2007). Petal abscission in rose (Rosa bourboniana var Gruss an Teplitz) is associated with the enhanced expression of an alpha expansin gene, RbEXPA1. *Plant Sci., 172*, 481-487.

Schroder, J., Stenger, H. & Wernicke, W. (2001). α-Tubulin genes are differentially expressed during leaf cell development in barley (Horgeum vulgare L.). *Plant Mol. Biol.*, *45*, 723-730.

Schnabl, H. & Mayer, I. (1976). Dark fixation of CO2 by flowers of cut roses. *Planta*, *131*, 51-55.

Shibaoka, H. (1994). Plant hormone-induced changes in the orientation of cortical microtubules: Alterations in the crosslinking between microtubules and the plasma membrane. *Annu. Rev. Plant Physiol. Plant Mol. Biol.*, *45*, 527-544.

Shimizu-Yumoto, K. & Ichimura, K. (2007). Effect of relative humidity and sucrose concentration on leaf injury and vase-life during sucrose pulse treatment in cut Eustoma flowers. *Hort. Research (Japan)*, *6*, 301-305.

Steen, D. A. & Chadwick, A. V. (1981). Ethylene effects in pea stem tissue. Evidence of microtubule mediation. *Plant Physiol.*, *67*, 460-466.

Takahashi, R., Fujitani, C., Yamaki, S. & Yamada, K. (2007). Analysis of the cell wall loosening proteins during rose flower opening. *Acta Hort.*, *755*, 483-488.

Tan, H., Liu, X., Ma, N., Xue, J., Lu, W., Bai, J. & Gao, J. (2006). Ethylene-influenced flower opening and expression of genes encoding Etrs, Ctrs, and Ein3s in two rose cultivars. *Postharv. Biol. Technol.*, *40*, 97-105.

Tang, G. Q., Lüscher, M. & Sturm, A. (1999). Antisense repression of vacuolar and cell wall invertase in transgenic carrot alters early plant development and sucrose partitioning. *Plant Cell*, *11*, 177-189.

Trouverie, J., Chateau-Joubert, S., Thévenot C., Jacquemot, M. P. & Prioul, J. L. (2004). Regulation of vacuolar invertase by abscisic acid or glucose in leaves and roots from maize plantlets. *Planta*, *219*, 894-905.

Uehara, U. & Hogetsu, T. (1993). Arrangement of cortical microtubules during formation of bordered pit in the tracheids of Texus. *Protoplasma*, *172*, 145-153.

van Doorn, W. G., Schurer, K. & Witte, Y. D. (1989). Role of endogenous bacteria in vascular blockage of cut rose flowers. *J. Plant Physiol.*, *134*, 375-381.

van Doorn, W. G., Groenewegen, G., van de Pol, P. A. & Berkholst, C. E. M. (1991). Effects of carbohydrates and water status on opening of cut Madelone roses. *Postharv. Biol. and Technol.*, *1*, 47-57.

van Doorn, W. G. & D'hont, K. (1994). Interaction between the effects of bacteria and dry storage on the opening and water relations of cut rose flowers. *J. Appl. Bacteriol.*, *77*, 644-649.

van Doorn, W. G. (1997). Water relations of cut flowers. *Hort. Rev.*, *18*, 1-85.

van Doorn, W. G. & van Meeteren, U. (2003). Flower opening and closure: a review. *J. Exp. Bot.*, *54*, 1801-1812.

van Doorn, W. G. & Kamdee, C. (2014). Flower opening and closure: an update. *J. Exp. Bot.*, *65*, 5749-57.

Vergauwen, R., van den Ende, W. & van Laere, A. (2000). The role of fructan in flowering of Campanula rapunculoides. *J. Exp. Bot.*, *51*, 1261-1266.

Vissenberg, K., Martinez-Vilchez, I. M., Verbelen, J.-P., Miller, J. G. & Fry, S. C. (2000). *In vivo* colocalization of xyloglucan endotransglycosylase activity and its donor substrate in the elongation zone of Arabidopsis roots. *Plant Cell*, *12*, 1229-1238.

Wagner, G. J. (1979). Content and vacuole/extravacuole distribution of neutral sugars, free amino acids, and anthocyanin in protoplasts. *Plant Physiol.*, *64*, 88-89.

Wang, M., Ding, L., Gao, L., Li, Y., Shen, Q. & Guo, S. (2016). The Interactions of Aquaporins and Mineral Nutrients in Higher Plants. *Int. J. Mol. Sci.*, *17*, 1229.

Welbaum, G. E. & Meinzer, F. C. (1990). Compartmentation of solutes and water in developing sugarcane stalk tissue. *Plant Physiol.*, *93*, 1147-1153.

Wernett, H. C., Wilfret, G. J., Sheehan, T. J., Marousky, F. J., Lyrene, P. M. & Knauft, D. A. (1996). Postharvest longevity of cut-flower

Gerbera. I. Response to selection for vase life of components. *J. Amer. Soc. Hort. Sci.*, *121*, 216-221.

Wernicke, W., Gunther, P. & Jung, G. (1993). Microtubules and cell shaping in the mesophyll of *Nigella damascena* L. *Protoplasma*, *173*, 8-12.

Wu, M. J., van Doorn, W. G. & Reid, M. S. (1991). Variation in the senescence of carnation (Dianthus caryophyllus L.) cultivars. I. Comparison of flower life, respiration and ethylene biosynthesis. *Sci. Hortic.*, *48*, 99-107.

Yamada, K., Ito, M., Oyama, T., Nakada, M., Maesaka, M. & Yamaki, S. (2007). Analysis of sucrose metabolism during petal growth of cut roses. *Postharvest Biol. Technol.*, *43*, 174-177.

Yamada, K., Norikoshi, R., Suzuki, K., Imanishi, H. & Ichimura, K. (2009a). Determination of subcellular concentrations of soluble carbohydrates in rose petals during opening by nonaqueous fractionation method combined with infiltration-centrifugation method. *Planta*, *230*, 1115-1127.

Yamada, K., Takahashi, R., Fujitani, C., Mishima, K., Yoshida, M., Daryl, C. J. & Yamaki, S. (2009b). Cell wall extensibility and effect of cell-wall-loosening proteins during rose flower opening. *J. Japan. Soc. Hort. Sci.*, *78*, 242-251.

Yamada, K., Norikoshi, R., Suzuki, K., Nishijima, T., Imanishi, H. & Ichimura, K. (2009c). Cell division and expansion growth during rose petal development. *J. Japan. Soc. Hort. Sci.*, *78*, 356-362.

Zenoni, S., Reale, L., Tornielli, G. B., Lanfaloni, L., Porceddu, A., Ferrarini, A., Moretti, C., Zamboni, A., Speghini, A., Ferranti, F. & Pezzotti, M. (2004). Downregulation of the Petunia hybrida α-expansin gene PhEXP1 reduces the amount of crystalline cellulose in cell walls and leads to phenotypic changes in petal limbs. *Plant Cell*, *16*, 295-308.

Zhang, L. Y., Peng, Y. B., Pelleschi-Travier, S., Fan, Y., Lu, Y. F., Lu, Y. M., Gao, X. P., Shen, Y. Y., Delrot, S. & Zhang, D. P. (2004). Evidence for apoplasmic phloem unloading in developing apple fruit. *Plant Physiol.*, *135*, 574-586.

In: Agricultural Research Updates. Vol. 28 ISBN: 978-1-53616-253-0
Editors: P. Gorawala, S. Mandhatri

Chapter 3

EFFECT OF PROCESSING OPERATIONS ON CARROTS' PROPERTIES

Raquel P. F. Guiné[1,*] and Sofia G. Florença[2]
[1]CI&DETS and CERNAS Research Centers,
Polytechnic Institute of Viseu, Viseu, Portugal
[2]Faculty of Nutrition Sciences, University of Porto, Oporto, Portugal

ABSTRACT

Carrots are rich in many bioactive molecules, which provide health benefits for the human body. Among these are carotenoids, vitamins, saccharides or phenolic compounds like anthocyanins and others. However, it is well known that when carrots are ingested fresh they provide the highest bioactivity, and the processing operations such as boiling, frying, freezing, drying, among others, induce some chemical and structural changes that may diminish the bioactivity of the compounds present. Furthermore, pre-treatment operations, like cutting or bleaching are also known to influence the quality of the food products. For example, the thermal pasteurization of carrot juice influences some vitamins and carotenoids. Also, the drying of carrots to incorporate into dried formulae

* Corresponding Author's E-mail: raquelguine@esav.ipv.pt.

greatly imparts the anthocyanins' content. Hence, the objective of the present chapter was to present some recent findings about the influence of processing operations on the properties and bioactive compounds in carrots. To achieve this purpose, different pre-treatments and processing operations are discussed as to their effects on some typical carrot products.

Keywords: bioactive compounds, carrots, pre-treatment, processing

1. INTRODUCTION

Carrot (*Daucus carota* L.) is a biennial root vegetable belonging to the Apiaceae family that produces leaves and a root, most frequently orange, although other colors may also apply, such as purple, black, red, white or yellow (D'Hooghe et al., 2018).

The carrot species cultivated in the present are a domesticated form of the wild carrot, native to Europe, North Africa, and Asia. It is assumed that the plant might have appeared firstly in Persia and was originally cultivated for the leaves as well as the seeds. In the present day, taproot is the most representative part of the plant consumed, although the stems and leaves are also eaten in a lower extension. The domestic carrot has resulted from selective breed, according to the desirable quality criteria, namely larger size, more palatable taste and smoother texture (Mueller-Bieniek, 2010).

Carrot crops are sensitive to agricultural practices that determine soil water-holding capacity, nutrient retention as well as soil aeration and infiltration rates. Hence, root yield and quality may be affected by such management operations. Nevertheless, the size is also highly affected by the initial distribution of the seeds on the soil, which defines the environment for the growth of the parts below ground. Also, the growth of the roots may be constrained by factors of physical nature, like, for example, high soil bulk density or the presence of stones. Nutrient availability and neighboring plant competition also greatly impart the size of the roots and crop yield. The growth of carrots starts with the development of a rosette of leaves above ground while the taproot builds up in the underground. The time between sowing and harvesting may vary between 3 months (~90 days), for the fast growing cultivars, and 4 months (~120 days) for the slower-maturing cultivars (D'Hooghe et al., 2018; Lana, 2012).

Carrot roots are consumed raw, in salads, or cooked as ingredient in may gastronomic preparations. Other applications include processing into pickles and juices. On the other hand, the seeds are used for extraction of essential oil (D'Hooghe et al., 2018; Nagachandrabose, 2018).

Carrot is undoubtedly one of the most important vegetables, consumed around the world in huge quantities. Its culture involves extensive areas and represents an important part of world food trade, with great socioeconomic involvement of farmers, processors and traders. In economic terms, carrot is one of the most important vegetables worldwide, and can be consumed raw or in a wide variety of processed products. On average, the worldwide carrot consumption is around 4.3 kg per capita per year, thus demonstrating its place as one of the most economically valuable root vegetables (Cubero-Leon et al., 2018; Freitas et al., 2009; Souza et al., 2019).

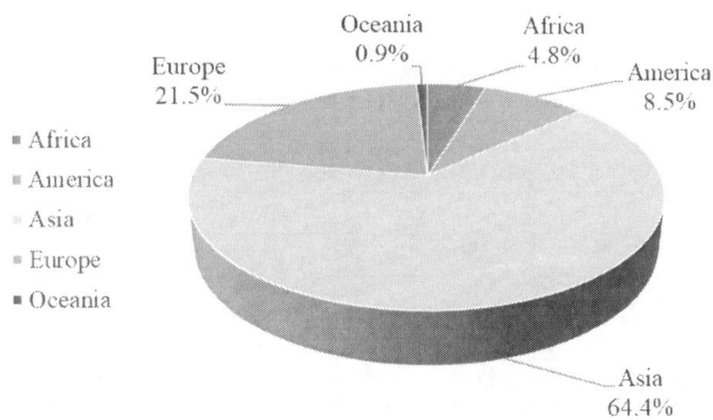

Figure 1. Distribution by continent of the worldwide carrot production in the year 2017 (FAOSTAT, 2017).

Carrot is one of the top ten vegetable crops grown worldwide in tropical and subtropical areas. The United Nations Food and Agriculture Organization (FAO) reports that, for the year 2017, the world cultivated area for carrots and turnips (these two crops appear combined in the FAOSTAT database) was 1.1 million ha, corresponding to a global production of 42.8 million tonnes (FAOSTAT, 2017). Figure 1 shows the distribution by continent of the worldwide carrot production.

Table 1 presents the areas for the carrot and turnip crops, together with the yield and quantities produced in the year 2017, globally and separated by world region (FAOSTAT, 2017).

Carrot is cultivated commercially in more than 130 countries, with China leading the list of the ten top carrot producers, representing almost 50%, as shown in Table 2 (FAOSTAT, 2017; Nagachandrabose, 2018).

Table 1. Statistics for carrot crop in the year 2017 (FAOSTAT, 2017)

Continent	Harvested area (Thousand ha)	Yield (Tonnes/ha)	Production (Thousand tonnes)
Africa	120	171	2,054
America	111	326	3,628
Asia	643	428	27,572
Europe	265	347	9,206
Oceania	8	486	372
World	1,147	373	42,832

Table 2. Top ten carrot producers in the year 2017 (FAOSTAT, 2017)

Country	Production (Thousand tonnes)	Percentage (%)
China	20,274	47.3
Uzbekistan	2,250	5.3
Russian Federation	1,806	4.3
USA	1,540	3.6
UK	957	2.2
Ukraine	839	2.0
Poland	827	1.9
Germany	734	1.7
Turkey	571	1.3
Japan	571	1.3
Rest of the world	12,463	29.1
Total	42,832	100.0

2. Composition of Raw Carrots

Carrot roots constitute a rich source of vitamins: carotenoids (alpha and beta carotene, which are precursors of vitamin A), vitamin K and vitamin B_6, as well as minerals: potassium, manganese, sodium, iron, zinc, copper and calcium. However, the chemical composition of carrots is highly variable according to cultivar and growing practices, such as the use of fertilizers or pesticides (D'Hooghe et al., 2018; Nagachandrabose, 2018).

Table 3. Proximate composition of raw carrot (USDA, 2018)

Component (g/100 g)	Carrot	Baby carrot
Water	88.29	90.35
Protein	0.93	0.64
Total lipid (fat)	0.24	0.13
Saturated fatty acids	0.032	0.023
Monounsaturated fatty acids	0.012	0.007
Polyunsaturated fatty acids	0.102	0.065
Trans fatty acids	0.00	0.00
Cholesterol	0.00	0.00
Ash	0.97	—
Carbohydrate	9.58	8.24
Total dietary fibre	2.80	2.9
Starch	1.43	—
Total sugars	4.74	4.76
Sucrose	3.59	—
Glucose (Dextrose)	0.59	—
Fructose	0.55	—
Lactose	0.00	—
Maltose	0.00	—
Galactose	0.00	—
Energy (kJ)	41	35
Energy (kcal)	173	147

Tables 3 to 6 show the chemical composition of raw carrot based on the data available on the United States department of Agriculture Food

Composition databases (USDA, 2018), and can be taken as a reference. Carrots, like many other vegetables, are mainly composed of water, around 90% (Table 3), followed by carbohydrates, about 10%. Other proximate components are residual, like for example protein (<1%) or ash (~1%) and fat is practically absent. The chemical composition of regular orange carrots and baby carrots is similar, as the results in Table 3 demonstrate.

Table 4. Mineral composition of raw carrot (USDA, 2018)

Component	**Carrot**	**Baby carrot**
Calcium, Ca (mg/100 g)	33	32
Copper, Cu (mg/100 g)	0.045	—
Fluoride, F (µg/100 g)	3.2	—
Iron, Fe (mg/100 g)	0.30	0.89
Magnesium, Mg (mg/100 g)	12	10
Manganese, Mn (mg/100 g)	0.143	—
Phosphorus, P (mg/100 g)	35	28
Potassium, K (mg/100 g)	320	237
Selenium, Se (µg/100 g)	0.1	—
Sodium, Na (mg/100 g)	69	78
Zinc, Zn (mg/100 g)	0.24	0.17

From the carbohydrates present, about one third is dietary fiber (~3%), which is associated with health benefits, most especially when ingested trough fruits and vegetables owing to the interaction with other phytochemicals. Some of these effects include the ability to reduce cholesterol, triglycerides, systolic blood pressure, glycaemia and insulin sensitivity. Furthermore, it can help control body weight, enhance the immune system response, help control diabetes and prevent many disorders of the gastrointestinal system, such as gastroesophageal reflux, duodenal ulcers, irritable bowel syndrome, diverticulitis, constipation or hemorrhoids (Guiné et al., 2016a, 2016b; Ljubicic et al., 2017).

Table 4 shows that the most abundant dietary minerals in carrot are potassium, sodium, phosphorous and calcium. According to a study by Krejčová et al., (2016), production mode (organic *versus* conventional) influences the mineral composition of carrots, particularly nitrates, which were found in higher concentration in carrot juice obtained from conventional farming together with higher amounts of NH_3, Mg and Na. Conversely, Ca was higher in organic juices. The authors proposed explanations for these differences between crops cultivated under organic and conventional growing systems to the higher levels of *Arbuscular mycorrhizal* fungi growing in the soils of organic crop fields.

Table 5. Vitamin content in raw carrot (USDA, 2018)

Component	Carrot	Baby carrot
Alfa-carotene (μg/100 g)	3477	—
Beta-carotene (μg/100 g)	8285	—
Betaine (μg/100 g)	0.4	—
Choline, total (mg/100 g)	8.8	—
Lycopene (μg/100 g)	1	—
Lutein + zeaxanthin (g/100 g)	256	—
Vitamin A, retinol RAE1 (μg/100 g)	835	690
Vitamin C, total ascorbic acid (mg/100 g)	5.9	2.6
Vitamin B1, thiamin (mg/100 g)	0.066	0.030
Vitamin B2, riboflavin (mg/100 g)	0.058	0.036
Vitamin B3, niacin (mg/100 g)	0.983	0.556
Vitamin B5, pantothenic acid (mg/100 g)	0.273	—
Vitamin B6, pyridoxin (mg/100 g)	0.138	0.105
Vitamin B9, total folate (μg/100 g)	19	27
Vitamin B12, cobalamin (μg/100 g)	0.0	0.0
Vitamin D, D2 + D3 (μg/100 g)	0.0	0.0
Vitamin E, alpha-tocopherol (μg/100 g)	0.66	—
Vitamin K, phylloquinone (μg/100 g)	13.2	9.4

1 RAE = retinol activity equivalents.

Carrots are known to be rich in carotenoids, which are precursors of vitamin A, as well as vitamins C or K. Carrot is considered the major and cheapest dietary source of vitamin-A, with 3477 and 8285 μg/100 g of alpha- and beta-carotene, respectively and 835 μg/100 g of vitamin A, measured as retinol equivalents (Table 5) (Chen et al., 2018; Tabor et al.,

2016). Vitamin K and B_9 contents are also important, varying from 9.4 to 13.2 and from 19 to 27 μg/100 g (USDA, 2018).

Table 6. Amino acids in carrot (USDA, 2018)

Amino acid	Amount (g/100g)	Amino acid	Amount (g/100g)
Alanine (Ala)	0.113	Lysine (Lys)	0.101
Arginine (Arg)	0.091	Methionine (Met)	0.020
Aspartic acid (Asp)	0.190	Phenylalanine (Phe)	0.061
Cystine (Cys)	0.083	Proline (Pro)	0.054
Glutamic acid (Glu)	0.366	Serine (Ser)	0.054
Glycine (Gly)	0.047	Threonine (Thr)	0.191
Histidine (His)	0.040	Tryptophan (Trp)	0.012
Isoleucine (Ile)	0.077	Tyrosine (Tyr)	0.043
Leucine (Leu)	0.102	Valine (Val)	0.069

Although the protein content in carrots is low, the biological value is considerable due to the diversity of amino acids present, as shown in Table 6. The most abundant are Glu (0.366 g/100 g), Thr (0.191 g/100 g) and Asp (0.190 g/100 g) (USDA, 2018).

3. Carrot Processing Operations and Their Effects on Product Quality

3.1. Minimally Processed Carrot

In recent times, extensive effort has been dedicated by food processors to the development of innovative ready-to-eat (RTE) food products, whose characteristics are those of a fresh product. However, these have the convenience of being pre-prepared for consumption, thus matching modern urban lifestyles, where time for food preparation is limited. As a consequence, today's consumers can find a variety of products ranging from ready-to-heat cooked whole meals to freshly prepared refrigerated salad snacks. Many consumers recognize the benefits of diets rich in fruits and vegetables, containing micronutrients such as vitamins or phenolic compounds, and value uncooked minimally processed (MP) vegetables as a healthiest option. Carrots are often a key ingredient in RTE vegetable products, representing up to 20% of the total weight, and they can be present in variable forms, including shreds, slices, discs or cubes and batons (González-Aguilar et al., 2009; Trichopoulou et al., 2003).

Vegetable salad is frequently prepared with fresh lettuce, carrots, tomato, onion, cabbage, among other ingredients. However, the counterpart to the benefits associated with fresh vegetables' consumption relies on undesirable contaminants present, which can be of chemical, physical or microbiological nature (Ameko et al., 2012). In the urban and peri-urban centers of less developed countries or very high density populated areas, contaminated water is sometimes used to grow vegetables. Besides the irrigation water, contamination can also come from the application of manure or from the contaminated soil. Furthermore, the microflora on raw fruits and vegetables is dominated by spoilage bacteria, yeasts and molds, so that it is highly recommended an adequate treatment for decontamination of products irrigated with waste water before consumption (Tiimub et al., 2012; Timbilla and Nyarko, 2004).

3.1.1. Mechanical Operations

MP carrots contain bioactive aliphatic C17-polyacetylenes (falcarinol, falcarindiol, falcarindiol-3-acetate) in small amounts, whose retention was evaluated by Koidis et al., (2012) for carrots prepared in four different forms: disc cutting, baton cutting, cubing and shredding. The carrots were submitted to peeling, cutting and washing with chlorinated water and then stored for 7 days. The results showed that the initial unit operations influenced the polyacetylene retention, with particular emphasis to peeling, since in this operation the peels with high polyacetylene content were discarded. In most cases, when washing was performed after cutting, lower retention was observed possibly owing to leakage during tissue damage occurred in the cutting operation. On the other hand, during storage a relatively high retention was observed, indicating a high plant matrix stability. In conclusion, mechanical operations and especially abrasive peeling performed to MP carrot might need additional optimization in order to be milder and minimize losses of bioactive compounds.

3.1.2. High Pressure Processing

Because customers' demand for MP foods (vegetables and fruits) is increasing and customers are more conscientious about quality and safety issues, high pressure processing (HPP) is also increasing as one of the choice commercially utilized techniques in the food industry. Because HPP is not able to break the covalent bonds of the compounds that provide nutritional and organoleptic properties, the freshness of fruit or vegetable products submitted to HPP can be better retained. Furthermore, HPP leads to inactivation of microorganisms and endogenous enzymes, and therefore microbial spoilage and quality deterioration during shelf life are minimized (Gao et al., 2016; Medina-Meza et al., 2014; Ramos et al., 2013).

Sun et al., (2018) investigated the effect of HPP treatments with 2 minutes duration at pressures in the range 400–1200 MPa on the texture of carrot. HPP originated losses of 71%, 74% in hardness and chewiness, respectively, for treatments at 400 MPa, but no further losses were observed as pressure increased. However, for pressures over 1000 MPa cohesiveness and resilience were recovered. Globally, the results indicated that carrot texture was variable according to pressure-induced modifications on cell morphology, membrane integrity and pectin properties.

3.1.3. Ionizing Radiation

The effect of ionizing radiation (IR) on the microbiological quality of MP carrot was evaluated by Frimpong et al., (2015). For this, MP carrot samples were analyzed for total viable count, total coliform count and pathogenic organisms, along a period of 15 days after treatment. The results obtained indicated that the predominant pathogenic organisms identified were *Bacillus cereus*, *Cronobacter sakazakii*, *Staphylococcus aureus*, and *Klebsiella spp*. And that a dosage of 2 kGy was recommended for treatment of minimally processed carrot.

3.1.4. Gaseous Chlorine Dioxide Treatment

Chlorine dioxide (ClO_2) gas is a strong oxidizing and sanitizing agent with high penetration capacity and a wide and solid biocidal effectiveness. For this reason, Gómez-López et al., (2007) evaluated its efficacy to prolong the shelf-life of MP carrot. The carrot's characteristics analyzed during shelf-life were O_2 and CO_2 headspace concentrations, microbiological quality (mesophilic aerobic bacteria, psychrotrophs, lactic acid bacteria, and yeasts), sensory quality (odour, flavour, texture, overall visual quality, and white blushing), and pH. The results obtained showed that ClO_2 did not significantly affect the respiration rate of MP carrots, while at the same time diminishing the pH. The package used maintained O_2 headspace concentrations in treated samples in equilibrium and prevented CO_2 accumulation. After treatment, the initial sensory quality of MP carrots was maintained. A lag phase of at least 2 days was observed for mesophilic aerobic bacteria, psychrotrophs and lactic acid bacteria, while mesophilic aerobic bacteria and psychrotrophs increased parallelly. Although odor was an important attribute in sensory deterioration, it reached an unacceptable score only after the samples were already rejected from the microbiological point of view. Hence, the shelf-life extension was limited to one day due to the restricted effect of the ClO_2 treatment on yeast counts.

3.1.5. Use of Edible Coating

Packaging plays a noticeable role in preserving food quality, and adequate packaging is fundamental for the conservation and marketing phases. Antimicrobial films and coatings have emerged as innovative concepts of active packaging and have been developed to reduce, inhibit or delay the growth of microorganisms on the food surfaces. In most cases, microbial contamination occurs with a higher extension on the surface of the food, and for that reason an effective control of microbial growth in needed. Traditionally, antimicrobial agents are added directly to the foods, but their activity can be inhibited by many of the substances present in the food, thus limiting their efficiency. As an alternative, the use of antimicrobial films or coatings can be more efficient because these can selectively and gradually migrate from the package onto the food, providing a prolonged release and contributing to the maintenance of high concentrations along the whole shelf life (Appendini and Hotchkiss, 2002; Ouattara et al., 2000).

Durango et al., (2006) developed an edible antimicrobial coating based on a starch–chitosan matrix to test on MP carrot. The samples of MP carrot slices were immersed into several coatings, prepared with different concentrations of starch, glycerol and chitosan. The obtained results demonstrated that the presence of 1.5% chitosan in the coatings inhibited the growth of total coliforms and lactic acid bacteria throughout the 15 days' storage period at 10ºC. Hence, edible antimicrobial yam starch and chitosan coating appears to be a viable way to control microbiological growth in MP carrot.

3.2. Cooked Carrot

Although carrots can be consumed raw, they are frequently cooked to prepare dishes, being boiled, roasted, fried, among others (Koç et al., 2017).

Cooking processes produce changes in the chemical composition and in the physical structure of foods, which are variable according to temperature, pressure and time. Cooking may lead to solubilization of minerals and vitamins into the cooking media, as well as degradation of vitamins or proteins due to heat and also changes in fatty acid composition owing to lipid oxidation. In traditional cooking, vegetables are usually immersed in boiling water at atmospheric pressure and approximately 100ºC, but this type of boiling results in loss of flavor, color and vitamins due to the high temperature applied. As a consequence, alternative methods have been tested to cook foods at lower temperatures and for shorter periods, such as for example microwave heating, radio-waves, ohmic heating, vacuum treatments and high-pressure processing (Araya et al., 2009; Iborra-Bernad et al., 2013).

In a study developed by Koç et al., (2017) the physicochemical properties of carrot were evaluated when submitted to three cooking methods: traditional cooking (TC), sous-vide (SV) and cook-vide (CV). While for TC at 100ºC and atmospheric pressure the processing times were between 15 and 60 min, SV and CV were performed at 60–90°C for times between 90 and 150 min and between 20 and 60 min, respectively. Antioxidant activity, total phenolic compounds and vitamin C analyses revealed that degradation in SV was approximately half of that of CV and TC. Regarding color, it was very similar in all three samples. Therefore, the work allowed concluding the TC had more adverse effects on the nutritional characteristics of carrots.

3.3. Processed Carrot

In addition to fresh consumption, carrots are processed, incorporated as ingredients in foods, frozen, canned, turned into juices or prepared in the dried form, like chips, powders or cubes (Guiné et al., 2014).

3.3.1. Blanching

Carrots are typically blanched in order to inactivate endogenous enzymes, which are responsible for the oxidation of nutrients and for the production of off-flavor compounds. Furthermore, blanching also allows to eliminate the raw/green flavor, as well as to stabilize color and to prevent microbial growth. Firmness is one of the pivotal sensory attributes valued by consumers when appreciating carrot products. However, an investigation by Belie et al., (2002) showed that consumers prefer the textural properties of carrots submitted to low temperature and short duration blanching, i.e., they would expect the texture of processed carrot products to be as similar as possible to that of fresh carrots. Consequently, various pre-treatment strategies have been developed more recently in order to better to preserve the hardness of carrots during blanching (Leong et al., 2018; Rico et al., 2007).

Conventional blanching typically uses heated water or steam, and high temperatures are necessary to inactivate the enzymes which could adversely impact the product's properties, like for example color, texture or nutritional value. Besides, blanching with hot water traditionally generates high amounts of effluents and therefore alternative techniques have emerged. Dry-blanching does not involve addition of hot water or steam and includes treatments like infrared or microwave treatments (Srimagal et al., 2017).

3.3.1.1. Blanching by Radio Frequency

Radio frequency (RF) heating is a volumetric method, in which heat is generated in the food by mechanisms of ionic conduction and dipole rotation. Due to its deep penetration, it generates heat inside foods much faster than heating based on diffusion and transfer phenomena (Guiné, 2013). RF has been proposed as a dry-blanching method in recent studies, and Gong et al., (2019) investigated its application to carrot cubes, to evaluate the capacity to inactivate peroxidase as well as to study the effect on product quality in terms of texture, color and vitamin C content. Their results showed that radio frequency heating reduced peroxidase activity by 90–95% after a blanching treatment applied for 3 to 7 minutes. Furthermore, dry-blanching by RF had a minimum impact on texture, color and vitamin C content, contrarily to a control blanching treatment with hot water blanching.

3.3.1.2. Ultrasound Blanching

The sensory properties and the vitamin C content have been evaluated by Gamboa-Santos et al., (2013) in dried carrots previously treated with ultrasound (US) as compared with conventional blanching. They observed that, while conventional blanching treatments at high temperature produced a retention of vitamin C between 40% to 85%, the carrots submitted to pre-treatments using US showed much lower vitamin C retention rates, thus indicating the possible adverse effect of US in the vitamin C content.

3.3.1.3. Blanching by Pulsed Electric Field

Pulsed Electric Fields (PEF) is a mild processing technology that increases the permeability of cell membranes by means of electroporation of the cells in plant tissues. This is achieved by placing the products between two electrodes, which provide very short (ranging from milli- to micro-seconds) high voltage pulses. By adequately controlling PEF processing conditions, the cell permeability can be used to assist the transport of compounds of interest across the membranes. Such processes include, for example, eliminating water in dehydration operations, removing compounds in extraction processes, introducing molecules of nutritional or technological interest into the cells, in order to achieve food quality improvements (Dermesonlouoglou et al., 2016; Phoon et al., 2008; Toepfl et al., 2005).

Leong et al., (2018) investigated the possibility of using PEF to accelerate the infusion of calcium ions into carrot, as a means to preserve the textural characteristics, and in particular hardness, after blanching. Their results showed that calcium chloride infusion ($CaCl_2$) combined with PEF treatment importantly improved firmness of blanched carrots, around 60%, for shorter soaking time and lower calcium concentration. Furthermore, the authors further investigated the *in vitro* bioaccessibility of beta-carotene, and they found it was similar in the samples with PEF-enhanced calcium infusion and conventional calcium infusion.

3.3.2. Pre-Drying Operations

The appearance of dark pigments in foods as a consequence of processing is very frequent, and these phenomena, denominated as browning, result from Maillard reactions or from ascorbic acid-mediated process of caramelization. Nevertheless, the Maillard reactions explain most of the browning occurring in the drying of food products (Guiné et al., 2014).

It has been suggested that pre-drying treatment of solid foods can improve product quality and at the same time increase the mass transfer of moisture during drying by modifying the structure of the food products. These pre-treatments can be of physical or chemical nature and aim to inactivate enzymes such as polyphenoloxidase, peroxidase and phenolase, and also to inhibit some undesirable chemical reactions, such as Maillard reaction and caramelization, which many adversely affect the product's quality. Examples of pre-drying treatments include blanching, osmotic dewatering, sulfiting and immersing in aqueous solutions of calcium chloride, ascorbic acid, citric acid, and gelatinised starch. In addition, other operations such as high-intensity ultrasound or radiofrequency have emerged as novel pre-treatment techniques (Guiné et al., 2014).

3.3.2.1. Treatment with Ascorbic Acid

Barroca et al., (2014) studied the pre-treatment of carrots prior to drying, using ascorbic acid aqueous solutions at concentrations of 0.25% and 1% and for treatment times of 60 and 90 minutes. After soaking, the excess of solution was removed gently with paper tissues, and the carrot slices were dried with hot air at 60ºC. Their results showed that the tested soaking conditions were not effective in reducing the browning reactions during drying. However, regarding textural properties, these were considerably influenced by the used pre-treatments.

3.3.2.2. Treatment with Sodium Metabisulphite

Guiné et al., (2014) evaluated how pre-treatments influenced the physical and chemical properties of carrots submitted to convective drying at 60ºC. Before drying, the carrot slices were dipped in solutions of water plus sodium metabisulphite at concentrations of 0.25% and 1%, for periods varying from 1 to 1½ hours. Their results revealed that the dried carrots had similar chemical composition, regardless of use of pre-treatments, in terms of moisture content, crude fiber, ashes and sugars (reducing, total and non-reducing sugars). Furthermore, the browning of the dried carrots seemed to be independent of the pre-treatment, and the same was observed for the textural properties, and hence the results indicated the different combinations used had no apparent improvement on the quality of the dried carrots.

3.3.2.3. Microwave Treatment

Microwaves (MW) are electromagnetic waves with frequencies varying in the range from 300 MHz to 300 GHz. These waves produce volumetric heating, i.e., heat developed from the energy that is absorbed from the microwaves, thus leading to quicker heating rates (Gowen et al., 2006). For this reason, it is frequently used in combination with drying operations as a way to fasten the processes and at the same time minimizing energy costs.

The study by Haq et al., (2018) evaluated the effect of MW on dehydration characteristics of carrot shreds at different times (1, 2 and 3 min). The MW treatment at 2450 MHz produced a decrease of overall dehydration time, which was, after 3 min of MW treatment followed by drying at 50, 60 and 70°C temperatures approximately 18, 35 and 48%, respectively, as compared with untreated carrots. Furthermore, effective moisture diffusivity was lower for the untreated samples and highest for 3 min treated carrot shreds, and the activation energy for drying decreased from 36.31 to 14.24 kJ/mol.

3.3.2.4. Ultrasound Treatment

Ultrasound (US) has attracted considerable interest in drying processes when used as a pretreatment, because it allows to develop soft targeted processes that are effective in obtaining high quality products. US is, by definition, the energy produced by sound waves, in frequencies above the human hearing in a range from 16 kHz up to MHz. Because this range is wide, it can be divided into high frequency/low power (in the MHz range) and low frequency/high power (in kHz range). The low frequency US is commonly used in food processing operations, like for example freezing, drying, defoaming, enzyme activation or inactivation (Xu et al., 2014)

Wang et al., (2018) studied the effect of low frequency ultrasound (LFUS) pretreatment preceding intermediate-wave infrared radiation (IW-IR) drying on some quality indicators of carrot slices. The results obtained revealed that LFUS pretreatment caused an intense decrease in drying time due to the disruption of cell structures generating micro-channels to facilitate moisture diffusion. Additionally, the pre-treated dried carrot slices exhibited higher beta-carotene content and rehydration ratio when compared to the control samples. Also the color was more similar to the fresh carrot and the aromatic and volatile organic compounds of dried carrot increased when the sonication pre-treatment was applied.

3.3.3. Drying

Because fresh carrots are highly susceptible to decay owing to their high moisture content and water activity, it is imperative to avoid massive waste losses. Dehydration promotes water loss to a high extent, thus considerably lowering water activity, being for that reason one of the most common food preservation technologies used to extend shelf life of food products (Calado et al., 2016; Cruz et al., 2015).

The study by Calado et al., (2016) investigated the properties of carrots dried by hot air in the temperature range from 50ºC to 70ºC. Their results showed that when the drying temperature was highest (70ºC) the chemical components were highly affected (moisture, protein, fiber, ash and sugars), as well as the textural properties (hardness, gumminess and chewiness) and porosity, being for that reason not recommended to perform the drying of carrots at high temperatures. The color showed small variations in redness and yellowness, but with increase in lightness, resulting in a slight discoloration.

A novel drying process combining ultrasound and vacuum dehydration was developed by Chen et al., (2016) as a way to shorten the drying time and improve the quality of dried carrot slices. For this study carrot slices were dried with ultrasonic vacuum (USV) drying and also with simple vacuum drying at 65°C and 75°C. The obtained results showed that the drying rate was significantly influenced by the type of drying used as well as by temperature. Compared with vacuum drying, USV drying resulted in a 41–53% decrease in the drying time. Moreover, the rehydration potential, nutritional value (retention of beta-carotene and ascorbic acid), color, and textural properties of USV-dried carrot slices were considerably improved as compared with the vacuum-dried samples. Additionally, lower energy consumption was used in the USV drying.

The effect of simultaneous infrared dry-blanching and dehydration on quality characteristics of carrot slices was investigated by Wu et al., (2014). In their study, the authors tested the effects of various processing parameters on carrot samples exposed to infrared (IR) radiation heating to obtain simultaneous infrared dry-blanching and dehydration (SIRDBD). The results obtained indicated that high surface temperature and/or lower slice thickness produced faster enzyme inactivation as well as faster moisture removal. The SIRDBD process induced 1 log reduction in peroxidase activity, a moisture reduction of 88.8 g/100 g, overall color difference equal to 5.13 and vitamin C retention of 77.34 g/100 g, being these values considerably better when compared to the control. In view of these findings, the investigators concluded that SIRDBD could be used as an alternative to produce high quality blanched and partially dehydrated carrots.

Fruit and vegetable snacks are traditionally produced by freeze drying (FD), which, nevertheless, has some drawbacks, like long processing time and low capacity, leading to high energy consumption and increased costs. Although hot air (HA) drying is widely used, in the case of snacks' production it is not admissible as an isolated technique, because the products are not crispy enough, not presenting the desirable color and texture. Also, it is time and energy consuming. However, HA drying can be combined with other drying methods to produce popular snacks. For example, sequential HA drying and microwave vacuum drying can be adopted to yield carrot chips (Prakash et al., 2004). Chen et al., (2018) used sequential infrared blanching combined with HA drying to develop healthy crispy snacks from carrot. They used different drying systems: I) a solar cabinet drier, II) a fluidized bed drier (at temperatures 50, 60, 70°C) and III) a microwave oven drier (at power levels 2, 3, 4). The obtained results showed that carrots dried by fluidized bed presented better color, rehydration properties, greater beta-carotene retention and better sensorial acceptability when compared with the products obtained by microwave oven and solar drying operations.

Kroehnke et al., (2018) investigated the influence of airborne ultrasound-assisted convective drying and microwave-assisted convective drying, as well as their combination, on quality parameters of dried carrots, namely total color change, water activity, content of carotenoids, polyphenolic compounds and antioxidant capacity. Their results indicated that the use US and MW in convective drying allowed to decrease drying time by 9 to 81%, in which the highest reduction was obtained for simultaneous use of convection, ultrasound and microwave. Furthermore, it was also observed a product improvement when using ultrasound as compared to convective drying or microwave-convective drying.

3.3.4. Carrot Purée

The incorporation of powdered vegetables in a variety of foods allows to produce better nutritional foods and at the same time increase vegetable intake. The production of these fruit and vegetable purées mainly involves tissue disintegration, like blending or high pressure homogenization, followed by preservation, for example through thermal processing. These purées are complex multi-phase food systems consisting of particles dispersed in a continuous liquid phase (Santiago et al., 2016).

3.3.4.1. Pre-Treatments for Carrot Purée Preparation

The effect of blanching at low and high temperatures, as well as that of mechanical disruption on the flow properties of carrot purée was investigated by Christiaens et al., (2012). The results obtained revealed that the purée prepared from non-pretreated carrots presented high consistency and syneresis. On the other hand, pre-treatments that solubilize pectin, like high-pressure homogenization or high-temperature blanching, restricted syneresis. High-pressure homogenization was found useful to diminish particle size thus resulting in a lower consistency carrot purée.

Aguiló-Aguayo et al., (2014) tested pulsed electric fields (PEF) for the pre-treatment of carrot purees as a way to enhance the contents of carotenoids, polyacetylene and sugars and also improve overall color. Their results confirmed that the use of PEF pre-treatment (at 0.25 kV/cm for 10 ms (7.5 J/kg) increased redness (values of the color coordinate a*) and this was strongly correlated with the total carotenoid content. Furthermore, they found that PEF-treated samples had higher amounts of falcarindiol and falcarindiol-3-acetate as compared with unprocessed purées, but the extension of this increase was dependent on the operating conditions, namely treatment time. Also sucrose and β-glucose showed higher concentrations in PEF treated samples, again depending on the power applied and processing time. As a conclusion of their work, the authors admitted that PEF pre-treatment seems an appropriate process to increase the polyacetylene and sugar content in carrot purees, turning the product into a more nutritive one and with health benefits.

3.3.4.2. Carrot Purée Properties

Herbig et al., (2017) evaluated the stability of vitamin C and 5-methyltetrahydrofolate in carrot purée that had been re-heated by microwave, by Actifry® device and in a canteen water bath. This later showed good vitamin C retention, ensuring stability for 3 hours at 80ºC. Although vitamin C and 5-methyltetrahydrofolate are usually referred to as heat- and oxygen-susceptible molecules, the authors of the study observed that the three re-heating methods tested allowed good retention of these components, even when the surface was exposed to oxygen.

The changes in the concentration of beta-carotene and its bioaccessibility were evaluated by Knockaert e al. (2012) during carrot purée processing, as a function of operating conditions, namely addition of olive oil, high pressure (HP) homogenization, and thermal or high pressure pasteurization. The results of their research highlighted that HP homogenization enhanced the bioaccessibility of beta-carotene for treatments over 50 MPa, and this was further increased with later thermal pasteurization. However, the high temperature also produced some undesirable cis-isomers. On the contrary, HP pasteurization was not found to exert a positively effect on the concentration or bioaccessibility of beta-carotene, and further induced its oxidation.

Sharma et al., (2017) studied the rheological and sensory characteristics of pureed carrots, and they found that different hydrocolloids produced changes in structure and texture, although the sensory characteristics were very similar.

The chemical changes during storage of carrot purées submitted to thermal and high-pressure high-temperature (HPHT) sterilization were studied in the work by Kebede et al., (2015). They used some compounds as markers, namely Strecker aldehydes, terpenes, phenylpropanoids, fatty acid derivatives and carotenoids, because these degrade producing volatile compounds that change during shelf-life. While immediately after processing, the thermally sterilized samples showed higher amounts of Strecker aldehydes, the HPHT treated samples revealed higher quantities of naturally occurring terpenes and phenylpropanoids, which were better preserved with the treatments applied. However, by the end of storage, the concentration of these compounds decreased similar concentrations in both thermal and HPHT samples.

Leong et al., (2016) evaluated the effect of pulsed electric field (PEF) as a way to strengthen the bioprotective properties of carrot purées against oxidative damage induced by hydrogen peroxide (H_2O_2). The results of their research demonstrated that resistance towards oxidative damage was increased by PEF processing in carrot cultivars rich in carotenoids, as compared with untreated samples. On the other hand, for cultivars with low carotenoid levels, the application PEF effectively increased total carotenoid content, leading again to an improved bioprotective effect as compared with the untreated samples. Hence, the authors concluded that PEF could add value to carrot purées due to the maximization of the bioprotective effects.

Fermentation with lactic acid bacteria (LAB) constitutes a natural way to transform the technological and nutritional properties of foods and food ingredients. For this reason, the rheological, chemical and sensory properties of carrot purée were evaluated by Juvonen et al., (2015) as a function of fermentation with exopolysaccharide (EPS) producing LAB. The results obtained showed that from the six selected EPS producers, fermentations with five of those LAB affectively altered the texture of the purée, so that the formation of low-branched dextran was associated with thickness, while the production of beta-glucan was associated with elasticity. In this way, the work lead to the conclusion that fermentation with the selected EPS-producing LAB strains appears as a promising clean approach to substitute hydrocolloid additives which are used as texturizers in vegetable processed products.

3.3.5. Carrot Juice

Carrot juice is one of the most important processed carrot products. It is considered a healthy food, due to the high content in carotenoids, with pro-vitamin A activity and antioxidant properties. The carotenoids are situated in the chromoplasts of plant cells, and so their release from the raw material is limited by cell wall membranes, thus diminishing their bioaccessibility. Also, instability is an important drawback for cloudy juice, especially if freshly squeezed and not from concentrate. In this way, choosing appropriate mechanical processing techniques that reduce particle size and destroy structural barriers is an alternative to effectively improve juice stability and enhance the transfer of carotenoids from plant tissue into the juice (Palmero et al., 2016; Sharma et al., 2012).

3.3.5.1. High Pressure Homogenization

The influence of high pressure homogenization (HPH) on pectin structural characteristics and carotenoid bioaccessibility in carrot juice were investigated by Liu et al., (2019). The results obtained by these authors indicated that HPH improved the juice stability as well as carotenoid bioaccessibility and uronic acid content, while at the same time decreased the degree of methoxylation and acetylation of water-soluble pectin as compared to the control sample.

3.3.5.2. Ultraviolet Radiation

Riganakos et al., (2017) studied the preservation of carrot juice exposed to UV-C radiation and thermal treatment in terms of physicochemical, microbiological and sensorial characteristics. The results obtained in their experiments revealed that the shelf life of carrot juice samples increased from only 4 days in the fresh (control) sample to 12 days for the treated juices. Furthermore, it was observed that these treatments did not produce significant changes in the investigated juice properties, thus being suitable to extend the shelf life of carrot juice while maintaining good quality indicators (Riganakos et al., 2017).

3.3.5.3. Ultrasound

The effect of sonication and chemical treatment (with potassium metabisulfite, $K_2S_2O_5$,) on total phenolic compounds, flavonoids and antioxidant capacity of carrot juice during storage was studied by Nadeem et al., (2018). Additionally, the impact of sonication was evaluated for a storage period of 90 days under refrigeration. The researchers observed that sonication had a positive impact on nutritional properties of the juice, which showed higher total phenolic, flavonoid, reducing power and antioxidant properties.

3.3.5.4. High Pressure Processing

The effect of high-pressure processing (HPP) at 550 MPa for 6 min on carrot juice was studied by Zhang et al., (2016) as compared to high-temperature short-time (HTST) processing. For both treatments, the total plate count significantly decreased while yeasts and molds were completely inactivated. The juice submitted to HPP processing showed higher carotenoids, polyphenols, antioxidant capacity, viscosity, turbidity, and lower browning indices when compared with HTST-treated juice, while this later contained higher polyacetylenes. Additionally, HPP provided a more fresh-like quality and exhibited better aroma, taste, and overall acceptability. When the juices were stored for 20 days at 4°C, both were safe from the microbiological point of view, but their antioxidant properties decreased linearly due to a reduction in carotenoid and polyphenol contents and their organoleptic properties also deteriorated as storage time passed.

3.3.5.5. Probiotic Fermentation

Fruits and vegetables can be fermented with probiotics to further increase their functionality, thus adding value and greatly improving storage, including during transportation.

Studies have demonstrated that carrot juice fermented by *Lactobacillus plantarum* NCU 116, demonstrated better regulating properties for blood glycaemia, blood lipids, hormones and oxidative stress in type-II diabetic rats when compared with the non-fermented carrot juice. Furthermore, improved storage qualities were also observed for the fermented juice (Li et al., 2014; Wan et al., 2019; Xie, 2014).

Probiotic-fermented carrot juice is a functional food, and therefore colloidal stability during storage is essential to maintain the desirable properties. The water-soluble polysaccharides (WSP) present in carrot juice are pectinic, mainly containing galacturonic acid, galactose and arabinose as monosaccharide units, and they provide stability as colloidal stabilizers. In the work by Wan et al., (2019) the particle-size distribution, ζ potential, centrifugal sediment ratio, soluble solids content, total acid and structural features of these WSP were studied along storage time. The fermented carrot juice displayed pseudoplastic fluid behavior, and the molecular weight and solid morphology of WSPs remained practically unchanged along storage, although the molecular weights of WSPs decreased just slightly. Furthermore, viscosity and particle size remained relatively stable when stored at 4°C and 25°C, while centrifugal sediment ratio, soluble solids content and total acid all increased significantly. This can provide insights into the usage of ingredients which improve storage stability and other quality properties valued by consumers (Wan et al., 2019).

FINAL REMARKS

Processing has a relevant impact on the quality of vegetables and fruits as compared with the fresh counterparts. However, having in mind the advantages of processing in extending shelf-life or allowing diversification of the type of products consumed, some compromise in terms of quality might be acceptable. Modern technologies and processing operations have emerged in the past years to minimize loss of quality indicators, either at the chemical and nutritionals levels or also in terms of organoleptic acceptability, texture, color and shelf life extension.

Acknowledgments

The authors would like to acknowledge CI&DETS and CERNAS Research centers from Polytechnic Institute of Viseu, Portugal. The authors thank the valuable advice of the experts who revised the present chapter: Professor Maria João Barroca from Agrarian School of Polytechnic Institute of Coimbra, Portugal; Professor Luísa Paula Cruz-Lopes from Department of Environment, School of Technology and Management - IPV, CERNAS Research Centre, Viseu, Portugal.

References

Aguiló-Aguayo, I., Hossain, M. B., Brunton, N., Lyng, J., Valverde, J., and Rai, D. K. (2014). Pulsed electric fields pre-treatment of carrot purees to enhance their polyacetylene and sugar contents. *Innovative Food Science & Emerging Technologies 23*, 79–86.

Ameko, E., Achio, S., Alhassan, S., and Kassim, A. (2012). Microbial safety of raw mixed-vegetable salad sold as an accompaniment to street vended cooked rice in Accra, Ghana. *Afr J Biotechnol 11*, 11078–11085.

Appendini, P., and Hotchkiss, J. H. (2002). Review of antimicrobial food packaging. *Innovative Food Science and Emerging Technologies 3*, 113–126.

Araya, X. I. T., Smale, N., Zabaras, D., Winley, E., Forde, C., Stewart, C. M., and Mawson, A. J. (2009). Sensory perception and quality attributes of high pressure processed carrots in comparison to raw, sous-vide and cooked carrots. *Innovative Food Science & Emerging Technologies 10*, 420–433.

Barroca, M. J., Sério, S., Guiné, R. P. F., and Correia, P. M. R. (2014). Influence of pre drying treatment on physical properties of carrots. In *Proceedings of FoodBalt2014*, (Jelgava, Latvia), pp. 327–329.

Belie, N. D., Laustsen, A. M., Martens, M., Bro, R., and Baerdemaeker, J.D. (2002). Use of Physico-Chemical Methods for Assessment of Sensory Changes in Carrot Texture and Sweetness during Cooking. *Journal of Texture Studies 33*, 367–388.

Calado, A. R. P., Gonçalves, J. C., Correia, P. M. R., and Guiné, R. P. F. (2016). Time Evolution of Physicochemical Properties of Carrots During the Drying Process. *Current Biochemical Engineering 3*, 139–153.

Chen, J., Venkitasamy, C., Shen, Q., McHugh, T. H., Zhang, R., and Pan, Z. (2018). Development of healthy crispy carrot snacks using sequential infrared blanching and hot air drying method. *LWT 97*, 469–475.

Chen, Z.-G., Guo, X.-Y., and Wu, T. (2016). A novel dehydration technique for carrot slices implementing ultrasound and vacuum drying methods. *Ultrasonics Sonochemistry 30*, 28–34.

Christiaens, S., Van Buggenhout, S., Chaula, D., Moelants, K., David, C. C., Hofkens, J., Van Loey, A. M., and Hendrickx, M. E. (2012). In situ pectin engineering as a tool to tailor the consistency and syneresis of carrot purée. *Food Chemistry 133*, 146–155.

Cruz, A. C., Guiné, R. P. F., and Gonçalves, J. C. (2015). Drying Kinetics and Product Quality for Convective Drying of Apples (cvs. Golden Delicious and Granny Smith). *International Journal of Fruit Science 15*, 54–78.

Cubero-Leon, E., De Rudder, O., and Maquet, A. (2018). Metabolomics for organic food authentication: Results from a long-term field study in carrots. *Food Chemistry 239*, 760–770.

Dermesonlouoglou, E., Zachariou, I., Andreou, V., and Taoukis, P. S. (2016). Effect of pulsed electric fields on mass transfer and quality of osmotically dehydrated kiwifruit. *Food and Bioproducts Processing 100*, 535–544.

D'Hooghe, P., Diaz, D., Brunel-Muguet, S., Davy, M., Vial, F., Dubois, J., and Kauffmann, F. (2018). Spatial variation of root yield within cultivated carrot fields is strongly impacted by plant spacing. *Scientia Horticulturae 241*, 29–40.

Durango, A. M., Soares, N. F. F., and Andrade, N. J. (2006). Microbiological evaluation of an edible antimicrobial coating on minimally processed carrots. *Food Control 17*, 336–341.

FAOSTAT (2017). *FAOSTAT - Food and agriculture data.*

Freitas, F. C. L., Almeida, M. E. L., Negreiros, M. Z., Honorato, A. R. F., Mesquita, H. C., and Silva, S. V. O. F. (2009). Periods of Weed Interference in Carrot in Function of Spacing between Rows. *Planta Daninha 27*, 473–480.

Frimpong, G. K., Kottoh, I. D., Ofosu, D. O., and Larbi, D. (2015). Effect of gamma irradiation on microbial quality of minimally processed carrot and lettuce: A case study in Greater Accra region of Ghana. *Radiation Physics and Chemistry 110*, 12–16.

Gamboa-Santos, J., Cristina Soria, A., Pérez-Mateos, M., Carrasco, J. A., Montilla, A., and Villamiel, M. (2013). Vitamin C content and sensorial properties of dehydrated carrots blanched conventionally or by ultrasound. *Food Chemistry 136*, 782–788.

Gao, G., Ren, P., Cao, X., Yan, B., Liao, X., Sun, Z., and Wang, Y. (2016). Comparing quality changes of cupped strawberry treated by high hydrostatic pressure and thermal processing during storage. *Food and Bioproducts Processing 100*, 221–229.

Gómez-López, V. M., Devlieghere, F., Ragaert, P., and Debevere, J. (2007). Shelf-life extension of minimally processed carrots by gaseous chlorine dioxide. *International Journal of Food Microbiology 116*, 221–227.

Gong, C., Zhao, Y., Zhang, H., Yue, J., Miao, Y., and Jiao, S. (2019). Investigation of radio frequency heating as a dry-blanching method for carrot cubes. *Journal of Food Engineering 245*, 53–56.

González-Aguilar, G. A., Ayala-Zavala, J. F., Rosa, L. A. de la, and Alvarez-Parrilla, E. (2009). Phytochemical Changes in the Postharvest and Minimal Processing of Fresh Fruits and Vegetables. In *Fruit and Vegetable Phytochemicals*, (John Wiley & Sons, Ltd), pp. 309–339.

Gowen, A., Abu-Ghannam, N., Frias, J., and Oliveira, J. (2006). Optimisation of dehydration and rehydration properties of cooked chickpeas (Cicer arietinum L.) undergoing microwave-hot air combination drying. *Trends in Food Science and Technology 17*, 177–183.

Guiné, R. P. F. (2013). *Unit Operations for the Food Industry. Volume I: Thermal Processing & Nonconventional Technologies* (Germany: LAP Lambert Academic Publishing).

Guiné, R., Duarte, J., Ferreira, M., Correia, P., Leal, M., Rumbak, I., Baric, I., Komes, D., Satalic, Z., Saric, M., et al., (2016a). Knowledge about dietary fibres (KADF): development and validation of an evaluation instrument through structural equation modelling (SEM). *Public Health 138*, 108–118.

Guiné, R. P. F., Sério, S. I. A., Correia, P. M. R., and Barroca, M. J. (2014). Effect of pre-treatment on some physical-chemical properties of dried carrots. *Journal of Hygienic Engineering and* Design *6*, 187–191.

Guiné, R. P. F., Duarte, J., Ferreira, M., Correia, P., Leal, M., Rumbak, I., Barić, I. C., Komes, D., Satalić, Z., Sarić, M. M., et al., (2016b). Knowledge about sources of dietary fibres and health effects using a validated scale: a cross-country study. *Public Health 141*, 100–112.

Haq, R., Kumar, P., and Prasad, K. (2018). Effect of microwave treatment on dehydration kinetics and moisture diffusivity of Asiatic Himalayan black carrot. *Journal of the Saudi Society of Agricultural Sciences 17*, 463–470.

Herbig, A.-L., Mousties, C., and Renard, C. M. G. C. (2017). Impact of three warming-up methods on the stability of vitamin C and 5-methyltetrahydrofolate supplemented to apple and carrot purée. *LWT 84*, 668–673.

Iborra-Bernad, C., Philippon, D., García-Segovia, P., and Martínez-Monzó, J. (2013). Optimizing the texture and color of sous-vide and cook-vide green bean pods. *LWT - Food Science and Technology 51*, 507–513.

Juvonen, R., Honkapää, K., Maina, N. H., Shi, Q., Viljanen, K., Maaheimo, H., Virkki, L., Tenkanen, M., and Lantto, R. (2015). The impact of fermentation with exopolysaccharide producing lactic acid bacteria on rheological, chemical and sensory properties of pureed carrots (Daucus carota L.). *International Journal of Food Microbiology 207*, 109–118.

Kebede, B. T., Grauwet, T., Palmers, S., Michiels, C., Hendrickx, M., and Van Loey, A. (2015). Investigating chemical changes during shelf-life of thermal and high-pressure high-temperature sterilised carrot purees: A 'fingerprinting kinetics' approach. *Food Chemistry 185*, 119–126.

Knockaert, G., Lemmens, L., Van Buggenhout, S., Hendrickx, M., and Van Loey, A. (2012). Changes in β-carotene bioaccessibility and concentration during processing of carrot puree. *Food Chemistry 133*, 60–67.

Koç, M., Baysan, U., Devseren, E., Okut, D., Atak, Z., Karataş, H., and Kaymak-Ertekin, F. (2017). Effects of different cooking methods on the chemical and physical properties of carrots and green peas. *Innovative Food Science & Emerging Technologies 42*, 109–119.

Koidis, A., Rawson, A., Tuohy, M., and Brunton, N. (2012). Influence of unit operations on the levels of polyacetylenes in minimally processed carrots and parsnips: An industrial trial. *Food Chemistry 132*, 1406–1412.

Krejčová, A., Návesník, J., Jičínská, J., and Černohorský, T. (2016). An elemental analysis of conventionally, organically and self-grown carrots. *Food Chemistry 192*, 242–249.

Kroehnke, J., Szadzińska, J., Stasiak, M., Radziejewska-Kubzdela, E., Biegańska-Marecik, R., and Musielak, G. (2018). Ultrasound- and microwave-assisted convective drying of carrots – Process kinetics and product's quality analysis. *Ultrasonics Sonochemistry 48*, 249–258.

Lana, M. M. (2012). The effects of line spacing and harvest time on processing yield and root size of carrot for cenourete® production. *Horticultura Brasileira 30*, 304–311.

Leong, S. Y., Oey, I., and Burritt, D. J. (2016). Pulsed electric field improves the bioprotective capacity of purées for different coloured carrot cultivars against H2O2-induced oxidative damage. *Food Chemistry 196*, 654–664.

Leong, S. Y., Du, D., and Oey, I. (2018). Pulsed Electric Fields enhances calcium infusion for improving the hardness of blanched carrots. *Innovative Food Science & Emerging Technologies 47*, 46–55.

Li, C., Ding, Q., Nie, S.-P., Zhang, Y.-S., Xiong, T., and Xie, M.-Y. (2014). Carrot juice fermented with lactobacillus plantarum NCU116 ameliorates type 2 diabetes in rats. *Journal of Agricultural and Food Chemistry 62*, 11884–11891.

Liu, X., Liu, J., Bi, J., Yi, J., Peng, J., Ning, C., Wellala, C. K. D., and Zhang, B. (2019). Effects of high pressure homogenization on pectin structural characteristics and carotenoid bioaccessibility of carrot juice. *Carbohydrate Polymers 203*, 176–184.

Ljubicic, M., Saric, M. M., Rumbak, I., Baric, I. C., Komes, D., Satalic, Z., and Guiné, R. P. F. (2017). Knowledge about dietary fibre and its health benefits: A cross-sectional survey of 2536 residents from across Croatia. *Medical Hypotheses 105*, 25–31.

Medina-Meza, I. G., Barnaba, C., and Barbosa-Cánovas, G. V. (2014). Effects of high pressure processing on lipid oxidation: A review. *Innovative Food Science and Emerging Technologies 22*, 1–10.

Mueller-Bieniek, A. (2010). Carrot (Daucus carota L.) in Medieval Kraków (S. Poland): a cultivated form? *Journal of Archaeological Science 37*, 1725–1730.

Nadeem, M., Ubaid, N., Qureshi, T. M., Munir, M., and Mehmood, A. (2018). Effect of ultrasound and chemical treatment on total phenol, flavonoids and antioxidant properties on carrot-grape juice blend during storage. *Ultrasonics Sonochemistry 45*, 1–6.

Nagachandrabose, S. (2018). Liquid bioformulations for the management of root-knot nematode, Meloidogyne hapla that infects carrot. *Crop Protection 114*, 155–161.

Ouattara, B., Simard, R. E., Piette, G., Begin, A., and Holley, R. A. (2000). Inhibition of surface spoilage bacteria in processed meats by application of antimicrobial films prepared with chitosan. *International Journal of Food Microbiology 62*, 139–148.

Palmero, P., Panozzo, A., Colle, I., Chigwedere, C., Hendrickx, M., and Van Loey, A. (2016). Role of structural barriers for carotenoid bioaccessibility upon high pressure homogenization. *Food Chemistry 199*, 423–432.

Phoon, P. Y., Galindo, F. G., Vicente, A., and Dejmek, P. (2008). Pulsed electric field in combination with vacuum impregnation with trehalose improves the freezing tolerance of spinach leaves. *Journal of Food Engineering 88*, 144–148.

Prakash, S., Jha, S. K., and Datta, N. (2004). Performance evaluation of blanched carrots dried by three different driers. *Journal of Food Engineering 62*, 305–313.

Ramos, B., Miller, F. A., Brandão, T. R. S., Teixeira, P., and Silva, C. L. M. (2013). Fresh fruits and vegetables - An overview on applied methodologies to improve its quality and safety. *Innovative Food Science and Emerging Technologies 20*, 1–15.

Rico, D., Martín-Diana, A. B., Barat, J. M., and Barry-Ryan, C. (2007). Extending and measuring the quality of fresh-cut fruit and vegetables: a review. *Trends in Food Science and Technology 18*, 373–386.

Riganakos, K. A., Karabagias, I. K., Gertzou, I., and Stahl, M. (2017). Comparison of UV-C and thermal treatments for the preservation of carrot juice. *Innovative Food Science & Emerging Technologies 42*, 165–172.

Santiago, J. S. J., Christiaens, S., Van Loey, A. M., and Hendrickx, M. E. (2016). Deliberate processing of carrot purées entails tailored serum pectin structures. *Innovative Food Science & Emerging Technologies 33*, 515–523.

Sharma, K. D., Karki, S., Thakur, N. S., and Attri, S. (2012). Chemical composition, functional properties and processing of carrot-A review. *Journal of Food Science and Technology 49*, 22–32.

Sharma, M., Kristo, E., Corredig, M., and Duizer, L. (2017). Effect of hydrocolloid type on texture of pureed carrots: Rheological and sensory measures. *Food Hydrocolloids 63*, 478–487.

Souza, L. P., Faroni, L. R. D., Heleno, F. F., Pinto, F. G., Queiroz, M. E. L. R., and Prates, L. H. F. (2019). Difenoconazole and linuron dissipation kinetics in carrots under open-field conditions. *Ecotoxicology and Environmental Safety 168*, 479–485.

Srimagal, A., Mishra, S., and Pradhan, R. C. (2017). Effects of ethyl oleate and microwave blanching on drying kinetics of bitter gourd. *Journal of Food Science and Technology 54*, 1192–1198.

Sun, Y., Kang, X., Chen, F., Liao, X., and Hu, X. (2018). Mechanisms of carrot texture alteration induced by pure effect of high pressure processing. *Innovative Food Science & Emerging Technologies.*

Tabor, G., Yesuf, M., Haile, M., Kebede, G., and Tilahun, S. (2016). Performance of some Asian carrot (Daucus carota L. ssp. sativa Hoffm.) cultivars under Ethiopian conditions: Carrot and seed yields. *Scientia Horticulturae 207*, 176–182.

Tiimub, B. M., Kuffour, R. A., and Kwarteng, A. S. (2012). Bacterial contamination levels of lettuce irrigated with waste water in the Kumasi Metropolis. *J. Biol. Agric. Healthc. 2*, 116–127.

Timbilla, J. A., and Nyarko, K. O. (2004). A survey of cabbage production and constraints in Ghana. *Ghana Journal of Agricultural Science 37*, 93-101–101.

Toepfl, S., Heinz, V., and Knorr, D. (2005). Overview of pulsed electric field processing for food. In *Emerging Technologies for Food Processing*, pp. 69–97.

Trichopoulou, A., Naska, A., Antoniou, A., Friel, S., Trygg, K., and Turrini, A. (2003). Vegetable and fruit: The evidence in their favour and the public health perspective. *International Journal for Vitamin and Nutrition Research 73*, 63–69.

USDA (2018). *Food Composition Databases.*

Wan, Y.-J., Xu, M.-M., Gilbert, R. G., Yin, J.-Y., Huang, X.-J., Xiong, T., and Xie, M.-Y. (2019). Colloid chemistry approach to understand the storage stability of fermented carrot juice. *Food Hydrocolloids 89*, 623–630.

Wang, L., Xu, B., Wei, B., and Zeng, R. (2018). Low frequency ultrasound pretreatment of carrot slices: Effect on the moisture migration and quality attributes by intermediate-wave infrared radiation drying. *Ultrasonics Sonochemistry 40*, 619–628.

Wu, B., Pan, Z., Qu, W., Wang, B., Wang, J., and Ma, H. (2014). Effect of simultaneous infrared dry-blanching and dehydration on quality characteristics of carrot slices. *LWT - Food Science and Technology 57*, 90–98.

Xie, M. (2014). Mechanism and modeling of transcritical/supercritical fuel spray and mixture formation in internal combustion engines. Ranshao Kexue Yu Jishu/*Journal of Combustion Science and Technology 20*, 1–9.

Xu, B., Zhang, M., Bhandari, B., and Cheng, X. (2014). Influence of power ultrasound on ice nucleation of radish cylinders during ultrasound-assisted immersion freezing. *International Journal of Refrigeration 46*, 1–8.

Zhang, Y., Liu, X., Wang, Y., Zhao, F., Sun, Z., and Liao, X. (2016). Quality comparison of carrot juices processed by high-pressure processing and high-temperature short-time processing. *Innovative Food Science & Emerging Technologies 33*, 135–144.

Chapter Reviewed by:

Professor Luísa Paula Cruz-Lopes from Department of Environment, School of Technology and Management - IPV, CERNAS Research Centre, Viseu, Portugal.

Professor Dace Klava from Department of Food Technology, Latvia University of Life Sciences and Technologies, Jelgava, Latvia.

Professor Maria João Barroca from department of Food Science and Technology, Agrarian School of Polytechnic Institute of Coimbra, Portugal.

In: Agricultural Research Updates. Vol. 28 ISBN: 978-1-53616-253-0
Editors: P. Gorawala, S. Mandhatri

Chapter 4

IMPACT OF POSTHARVEST TECHNOLOGIES ON READY-TO-EAT SPINACH NUTRITIONAL QUALITY

Ana Oliveira*, Ana A. Vilas-Boas, Ana L. Amaro and Manuela Pintado
CBQF - Centro de Biotecnologia e Química Fina – Laboratório Associado, Escola Superior de Biotecnologia, Universidade Católica Portuguesa, Porto, Portugal

ABSTRACT

Spinach is an annual, green, leafy vegetable rich in health-promoting compounds, such as vitamins A and C, and flavonoids (Roberts and Moreau 2016). For consumer's acceptance, spinach leaves must be green and turgid. However, spinach undergo fast senescence processes after ripening and the quality attributes are quickly lost during postharvest handling (Oliveira et al. 2016). Inadequate postharvest management can result in significant nutritional losses due to the fact these leafy vegetables have high-water activity, neutral to slightly acidic pH values,

* Corresponding Author's E-mail: asoliveira@porto.ucp.pt.

nutritional complexity and can sustain the growth of many species of bacteria, yeasts and molds.

Spinach has relatively high respiration and water-loss rates, and is prone to tissue decay, microbial growth, and nutrients loss, all reducing the commercial potential, when not controlled. Spinach nutritional quality can be affected by postharvest technologies and storage conditions. Starting from spinach washing process until consumption, there is high potential for ascorbate and other nutrients loss, which are responsible for important bioactive properties such as antioxidant capacity (Esteban et al. 2019).

Recent postharvest technologies have been used to better maintain ready-to-eat spinach physiological and nutritional quality, namely the use of hydrogen peroxide, organic acids, gamma irradiation, UV, low irradiance pulses, ozone gas, intense pulsed light and a recently developed technology, the cold plasma, among others. In addition, packaging and storage conditions, focusing on temperature and relative humidity, are key factors affecting processed leafy vegetables quality, frequently used aiming to maintain high-value product quality throughout production, distribution and storage (Murray et al. 2017).

The knowledge and control of postharvest technologies impact on ready-to-eat spinach allows the attempt to modulate nutritional quality, bioactive properties and health-important compounds content increasing commercial value. This chapter discusses the theoretical aspects and recent developments in ready-to-eat spinach to improve preservation of nutritional quality after harvest and extend its shelf-life.

Keywords: fresh produce, ready-to-eat, *Spinacia oleracea* L., decontamination, postharvest quality

INTRODUCTION

Baby-sized leaves as ready-to-eat vegetable are one of the most promising developments in the sector of convenience food, being a natural source of health-promoting bioactives. The ready-to-eat products and their production technologies are evolving to provide more diverse commercial products with enhanced flavor and appealing characteristics (Saini, Ko, and Keum 2017).

From the field to the consumers' table, leafy vegetables must remain biologically active during processing, transport, and storage (both on supermarket shelves and consumers' refrigerators). A good nutritional quality is mandatory for consumers and should be maintained along product shelf-life (Esteban et al. 2019). Different steps of production, processing, storage, and transport are outlined in Figure 1.

Spinach (*Spinacia oleracea* L.) is one of the main leafy vegetables contributing to the ready-to-eat produce industry. Due to its tender and soft texture, spinach leaves tend to undergo different degrees of injuries during processing steps, which affect their quality and shelf life (Ariffin, Gkatzionis, and Bakalis 2017).

It is not possible to avoid quality losses by using a single technology. Instead, an integrated point of view of several strategies must be implemented. These include the sanitation during handling and processing, reduction of mechanical damage during processing, and treatments such as low temperature and low oxygen atmosphere packaging to reduce respiration rate and ethylene production (Ghidelli and Pérez-Gago 2018).

In the past, the postharvest wash step was considered an intervention with limited efficacy removing field-acquired contamination. The research performed focused on evaluating or formulating effective sanitizers. Research to enhance the microbiological safety of fresh produce has started to identify and develop alternative intervention methods. For example, irradiation, UV, ozone, pulsed light and chlorine dioxide have been reported as viable processes to decontaminate leafy vegetables surfaces together with those based on advanced oxidative processes (AOP) or gas plasmas, which are relatively recent advances (Murray et al. 2017).

Loosing product during postharvest represents high costs and means that the total investment in the production, harvest, processing and distribution are reduced without commercial return. Two of the most important variables that help maintain vegetables quality are storage temperature and time. There are recommended storage temperatures for fresh-cut baby spinach leaves; however, sometimes they are stored at higher temperatures, especially during transportation and limited accommodation in the markets. Low temperature (0–10ºC) storage helps reducing microbial spoilage, respiration and ethylene production rates, water loss and therefore the degradation of bioactive compounds in these products (Saini et al. 2016).

Other postharvest technologies such as controlled atmosphere (CA), modified atmosphere packaging (MAP) and edible coatings are supplementary to good temperature management. Packaging is important by improving the value of postharvest stage and delaying the quality loss during storage and transport of vegetables. Packaging conditions allows to control mechanical damage, but also to minimize the metabolic activity like respiration, transpiration and microbial growth (Mahajan et al. 2017). Package makes transportation, storage and dispensing of products more comfortable and allow product marketing (Wyrwa and Barska 2017; Otoni et al. 2017). MAP is an increasingly important type of packaging for ready-to-eat fruits and vegetables because it can reach lower respiration rate, delay ripening and discoloration, prevent the build-up of off-odors and inhibit the growth of pathogens and spoilage organisms (M. Zhang et al. 2015).

Until recently, control of light was not considered during postharvest storage and the products were commercially stored in darkness. Those conditions induced senescence by activating genes involved in chlorophyll, protein and chloroplast degradation, accelerating the production of reactive species of oxygen (Charles et al. 2018).

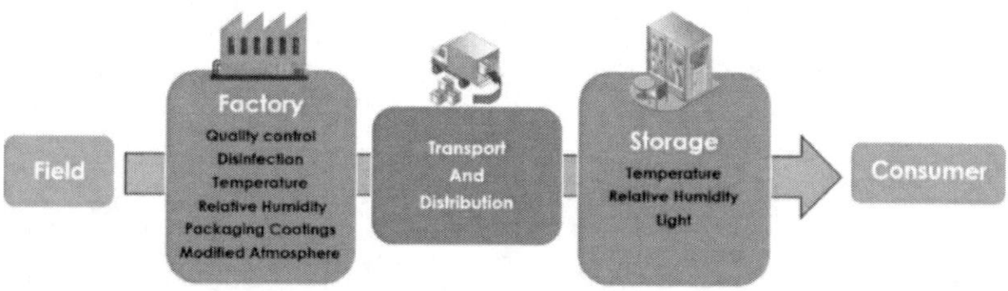

Figure 1. Different steps of processing, storage and transport in ready-to-eat spinaches.

General Procedures for Spinach Processing

Baby spinach leaves are mechanically harvested and then placed in trays, which may have ice to cool them down from the field heat when the temperature is above 27°C. After field harvest, the spinach are transported to the processing facilities where they are usually cooled by forced air systems. The minimum time and low temperature exposure occurring between harvests and processing are critical to minimize quality loss after harvest. When arriving to processing and packaging house, spinach is many times retained, with waiting times to get in the processing line. A recent study about the impact of temperature/RH during the waiting period between washing/drying and product packaging showed the importance of these periods conditions (Oliveira et al. 2016).

Spinach is manually loaded from containers onto an elevator conveyor, which transports the leaves and where they pass through an inspection belt, prior to entering the wash tanks. The washed spinach is then de-watered via centrifugation prior to being again transported and placed into packs and sealed. Throughout ready-to-eat processing line, spinach leaves may undergo several different technologies that ensure product safety while maintaining the fresh appearance. In general, leafy vegetables processing line includes inspection and selection, washing and sanitation, drying and packaging (Brasil and Siddiqui 2018). These different processing steps, using different technologies, and respective transitions not only affect safety and sensory quality but can also cause changes in nutritional quality and health-promoting properties of final product. Moreover, processing steps involving some mechanical damage promote faster microbial degradation, physiological alterations and biochemical changes as reactions between enzymes and substrates are enhanced by cells rupture and intercellular products leakage and ascorbic acid is oxidized once in contact with oxygen and light.

Postharvest Decontamination Technologies

Washing

Washing is one of the most used postharvest technologies as the removal of field contamination agents, such as soil and debris, irrigation water, insects and other animals, from the spinach leaf is of paramount importance to guarantee quality and safety throughout storage. Much attention has been given to the water quality maintenance during washing process in the attempt to maximize the washing effectiveness and minimize product crossover contamination. A diverse range of human pathogens has been associated with outbreaks linked to leafy vegetables although *Salmonella* and *Escherichia coli* O157:H7 are commonly the most implicated (Murray et al. 2017). These products can support the growth of many other different pathogenic bacteria, including *Shigella* and *Listeria*, because of their large surface area and potential for pathogen internalization in structures, such as the stomata.

Sanitizers, alone, or in combinations with other preservation strategies are commonly used in washing process to eliminate microorganisms from the spinach leaf surface, with chlorine as the worldwide most used sanitizer, due to its high efficacy and low cost. However, sodium hypochlorite solutions (chlorinated water) are considered to have low efficacy against specific bacterial communities and to be a potentially hazard carcinogenic compounds generator (Gu et al. 2018). There is, therefore, an effort to find health and environmentally friendlier strategies to replace the use of chlorinated water.

Ascorbate retention and degradation during the washing and postharvest storage of spinach showed that spinach leaves are more susceptible to loss of ascorbate during the washing process than wild rocket or watercress (Dewhirst et al. 2017). The loss of ascorbate would seem to be primarily due to mechanical stress, which manifests as oxidative stress, causing the oxidation of ascorbate. On the other hand, washing step with oxalic acid (1 mM for 1 min) improved the retention of antioxidant activity and total phenolic compounds content in spinach leaves during postharvest storage when compared to traditional sodium hypochlorite treatment (Cefola and Pace 2015). Also, single-step washing with citric acid (0.5%, 2.5 min, at 25ºC) represented an effective alternative to ready-to-eat spinach disinfection treatment (Finten, Agüero, and Jagus 2017). Although the organic acids, hydrogen peroxide and slightly acidic electrolyzed water use on washing process have been studied, the most recent investigation has been focused on the use of ozonated water (Karaca and Velioglu 2014; Papachristodoulou et al. 2018; Kang, Park, and Song 2019) and the use of quaternary ammonium surfactants (QAS).

Ozone, an approved sanitizing agent, is being tested widely for disinfection purposes in food industries. Due to its quick decomposition to oxygen that alleviates concerns about toxic residues, it has attracted much attention for food safety uses. Studies have verified that high microbial inactivation efficiency can be obtained with both aqueous and gaseous ozone treatments. However, due to its strong oxidizing activity it can cause detrimental effects on product quality. Karaca and Velioglu (2014) compared the efficiencies of distilled, ozonated (12.0 mg L^{-1}), and chlorinated (100 mg L^{-1}) water washes in inactivating *E. coli* and *L. innocua* and investigated the effects of these treatments on some nutritional characteristics (chlorophylls, ascorbic acid and total phenolic contents, and antioxidant activity) of spinach. Results showed that chlorine and ozone washes resulted in average log reductions of 2.9 and 2.0 for *E. coli*, respectively, while the efficiency of ozone (2.2 log) was very close to that of chlorine (2.3 log) on *L. innocua*. Moreover, aqueous ozone did not cause any detrimental effects on the nutritional value of spinach.

The effect of ozonated water was also evaluated at various concentrations (0, 0.4, 0.8, or 1.2 mg L^{-1}) and dipping times (0, 1, 15, or 30s), on the microbial counts and shelf-life attributes of ready-to-eat spinach. Results revealed that application of ozonated water (0.8 mg L^{-1} for 30 s) before packaging reduced yellowing and maintained compositional characteristics of ready-to-eat spinach leaves, ensuring a shelf-life extension of 3 days. Moreover, a positive effect on microbial population control was recorded during the first 5 days of storage (Papachristodoulou et al. 2018).

QAS are cationic surfactants, that can be used as sanitizers because they are environment-friendly, easy to handle, and non-corrosive (Meireles, Giaouris, and Simões 2016). Kang, Park, and Song (2019) determined the most effective QAS when used at low concentration as sanitizers for spinach against major foodborne pathogens, such as *E. coli* O157:H7, *Salmonella Typhimurium*, and *L. monocytogenes*. In addition, quality changes in spinach leaves after treatment with the chosen QAS were examined during storage at 4 °C. Spinach leaves inoculated with pathogenic bacteria were placed inside a plastic bath containing QAS (benzalkonium chloride, benzethonium chloride, cetylpyridinium chloride, cetyltrimethylammonium bromide, tetraethylammonium bromide) or NaOCl solution and then agitated for 3 min. The samples, after treatment, were dried for 30 min on a clean bench. Samples were packed in low density polyethylene bags and kept at 4 °C, for 6 days, to examine the quality changes in the samples during storage. Cetylpyridinium chloride (CPC) possessed the highest inhibitory activity among all the QAS tested, as well as NaOCl. It showed a 3.33–4.54 log-reduction against major foodborne pathogens inoculated on spinach leaves, when used at a concentration of 80 mg L^{-1}. CPC damaged and destroyed microbial cell membranes more effectively than NaOCl treatment, resulting in more release of cellular constituents. Additionally, deleterious effects on quality parameters of spinach were not observed after CPC treatment.

Irradiation

Gamma radiation is considered an effective mean of processing and preserving food products but literature reveals that radiation treatments have been shown to either increase or decrease the antioxidant content of foods, which is dependent on the dose delivered, exposure time, and the raw material used.

Irradiation (0.25 to 1.5 kGy gamma) application was studied in spinach bioactive compounds content and antioxidant activity. The results of this study reveal that gamma irradiation treatment significantly enhanced the spinach content of individual as well as total bioactive components and that positive correlation existed between gamma irradiation and total phenolics content (Hussain et al. 2016).

Ultraviolet irradiation (UV) has been used to prevent decay and to prolong the postharvest storage life of fruits and vegetables. It is well known that an appropriate dose of UV-C or UV-B treatment can greatly improve the maintenance of the quality of vegetables and fruits under the optimal conditions. The effects of UV-C treatment on the postharvest quality and biochemistry of spinach were studied by Liao's et al. (2016). Spinach was exposed to single and multiple UV-C irradiations at an energy level of 2.46 kJ m^{-2} and stored at 4ºC or 5 days. Samples were irradiated with germicide-emitting lamps located 20 cm above and below the surface, holding the trays. Each irradiation lasted for 5 min and the trays were rotated 90º every 2.5 min, to ensure uniform irradiation. For single UV-C irradiation, samples were lighted only once at the beginning. For multiple irradiations, the samples were treated once every day. The multiple UV-C treatments maintained significantly better quality parameters of postharvest spinach over the experiment, compared to the single treatment. Multiple UV-C treated spinach preserved better the contents of soluble protein, vitamin C and chlorophyll a.

UV-B treatment also increased the overall quality and visual scores of treated spinach. Ufuk Kasım and Kasım (2017) referred that with the highest level of UV-B tested (6 min; 0.9 kJ m^{-2}) better effects were obtained when compared with lower levels (2 min, 0.3 kJ m^{-2} 4 min, 0.6 kJ m^{-2}) and controls.

Intense Pulsed Light (IPL)

Pulsed light technology involves the use of inert-gas flash lamps to produce short-duration, high-power pulses of an intense broad spectrum of light within the frequency regions of ultraviolet (UV), visible (VL), and infrared (IR) light (200-1000 nm) (John and Ramaswamy, 2018). It represents an alternative to leafy vegetables thermal disinfection treatments as it is a postharvest technology applied to decontaminate food surfaces, by killing microorganisms using short time pulses of a broad spectrum, with a significant UV-C component. Its lethal action is attributed to a photochemical mechanism by which UV absorption produces severe damage in microbial DNA. Simultaneously, there is a photo thermal effect associated with the incident radiation-related overheating and a photophysical one, through the effect of light on membranes and cell constituents.

Agüero's et al. (2016) explored the suitability of intense pulsed light (IPL) treatments (20 and 40 kJ.m^2) to inactivate *Listeria innocua* and *Escherichia coli* inoculated on spinach and showed it was effective for both. IPL was also effective reducing initial total microbial counts showing high efficiency for coliforms. IPL treated spinach showed significantly increased polyphenolic content and antioxidant capacity (5–10% and 32–34%) for samples treated with 20 or 40 Kj m^{-2}, respectively. However, these authors also referred that treatment with IPL accelerated product metabolic rates, which resulted in faster decrease in these quality indicators during refrigerated storage.

The in-package application of IPL has potential as it allows treat the spinach when already in the package avoiding recontamination events but there are many factors to have in consideration and more research is needed for the guarantee of efficiency of in-package IPL treatment.

High-Pressure Processing (HPP)

High-Pressure Processing (HPP) is an innovative non-thermal postharvest technology based on the application of high pressure to the food product in flexible pouches. This high pressure application aims at the microbial cells inactivation, by inducing physical damage and protein denaturation. The optimized synergistic effect of pressure, time and temperature are reported to reduce total mesophilic aerobic bacteria, total yeasts and molds populations resulting in little to no effects on nutritional quality and sensory attributes as compared to thermal treatments. However, HPP is a batch process and in some cases, high pressure levels application for excessive time periods can result in negative effects on quality (e.g., increased lipid oxidation, texture changes) (Stratakos et al. 2015). To avoid the risks associated with high levels of pressure, and in the case of more sensible products as ready-to-eat leafy vegetables, HPP processing is combined with other strategies. Stratakos et al. (2016) studied the effect of

HPP (400 MPa min^{-1}), alone or in combination with *Weissella viridescens protective culture,* against *L. monocytogenes* in ready-to-eat salads with the pH values of 4.32 and 5.59, during storage at 4 and 12 °C. It was found that HPP at 400 MPa reduced the population of pathogens about 4.0 and 1.5 log CFU g^{-1}, at low and higher pH, respectively. The effect of HPP on *W. viridescens* growth in ready-to-eat salads stored at different temperatures (4 and 12°C) resulted in a small immediate reduction ranging from 0.50 to 1.2 log CFU g^{-1} depending on the applied pH value.

Gas Application

Once that the diffusivity of gases (~10^{-5} m^2 s^{-1}) is four orders of magnitude higher than liquids (~10^{-9} m^2 s^{-1}), gaseous sanitizers are usually able to penetrate within produce much faster than liquids (Shynkaryk et al. 2015). However, sanitizers such as chlorine dioxide or ozone are aggressive oxidizers, and may damage sensitive products such as spinach.

Chlorine dioxide (ClO_2) application has emerged as a postharvest non-thermal sanitizing technology because ClO_2 functions as a selective oxidant attacking cell membrane proteins and enzymes, increasing membrane permeability and penetrating the cell membrane damaging proteins and enzymes within the cell (Park and Kang 2017). However, the temperature at which the gas is applied has a great impact on the ClO_2 efficacy on foodborne pathogens inactivation. Park, Kang, and Kang (2018) showed that ClO_2 gas treatment at 15 °C caused significantly more and faster inactivation of the three pathogens studied than at 25 °C. The same authors, in the year before and using spinach as one of the models, showed that surface characteristics of produce and food contact surfaces have an influence on the inactivation of foodborne pathogens by ClO2 gas treatment (Park and Kang 2017). Surface hydrophobicity plays a more important role in bacterial inactivation on surfaces than does on surface roughness. Also, the existence of crevices with features of similar size to the pathogen cell was more important than the R_a and R_q values in the inactivation patterns of foodborne pathogens. Therefore, understanding surface characteristics is needed if this technology is to be commercialized.

Vacuum cooling causes bacterial internalization, making them harder to reach by sanitizer and the application of ozone during the vacuum cooling step significantly reduced population of viable internalized bacteria, which otherwise remain unaffected by sodium hypochlorite and UV light surface treatments. The efficacy of high ozone concentration (1.5 g m^{-3}) application during short term vacuum cooling step in combination with low ozone concentration (0.032 e 0.528 g m^{-3}) long term sanitization treatments (days) was also evaluated and the combination of gaseous treatment was found more effective in microbial reduction compared to a single treatment with a strongly expressed synergistic effect (Shynkaryk et al. 2016).

The efficacy of gaseous ozone application during vacuum cooling of spinach leaves was tested against *Escherichia coli* O157:H7 (Yesil et al. 2017). The study main goal was to evaluate the bacterium population size influence on the treatment efficacy but the study also shows that vacuum and ozone sequential treatment led to a greater than 3-log reduction of *E. coli* O157:H7 population. Inoculated leaves were vacuum cooled (28.5 in Hg; 4 °C) in a custom-made vessel and then subjected to a gaseous ozone treatment under the following conditions: 1.5 g of ozone per kg of gas mixture, vessel pressure at 10 lb in^{-2} gauge, 94 to 98% relative humidity, and 30 min of holding time at 9 °C. Besides the low number of studies regarding the application of gas to ready-to-eat spinach, there is a lack of information on the effect of these technologies on the nutritional quality of these products.

Cold Plasma Technology

Cold plasma is an emerging non-thermal processing technology that could be used for large scale leaf decontamination, as an alternative to chlorine washing. Cold plasma is produced by excitation of gas molecules through the use of electrical discharges. According to the frequency used to excite the gas, cold atmospheric plasma sources can be classified in direct current and low frequency, radio frequency and microwave discharges (Pasquali et al. 2016).

The strong electric field created by the device forms a discharge and produces a non-thermal plasma in which various active species such as hydroxyl radicals are produced. These radicals are expected to be extremely reactive.

Applying a high electric field to a stream of aerosolized sanitizer, such as hydrogen peroxide, using a specially designed aerosolizer, it was possible to use the sanitizer as a form of activated aerosol (Jiang et al. 2017). The efficacy of aerosolized hydrogen peroxide in inactivating bacteria and maintaining the quality of baby spinach leaves was investigated. *Escherichia coli* O157:H7, *Salmonella Typhimurium* and *Listeria innocua* inoculated spinach leaves were treated for 45 s followed by additional 30 min dwell time with hydrogen peroxide (7.8%) aerosols activated by atmospheric cold plasma. This study revealed that populations of *E. coli* O157:H7, *S. Typhimurium* and *L. innocua* inoculated on the smooth skin surface and stem scar area of spinach could be significantly reduced by a 45 s aerosolized H_2O_2 treatment plus 30 min dwell time.

Combination of Treatments

The ready-to-eat leafy vegetables industry, as others, applies various decontamination techniques throughout the processing chain, ranging from the chemical washing treatments to the emerging described technologies. Combinations of different technologies, known as hurdle technology, could be an alternative to the use of alone treatments with high levels or concentrations. Combined treatments could achieve required levels of food safety and the maintenance of nutritional quality of foods while decreasing the intensity of each hurdle, that is, the antimicrobial concentration.

Regarding the utilization of organic acids in washing treatment, a study where spinach leaves inoculated with *E. coli* (~7.00 log_{10} CFU g^{-1}) were treated with lactic acid solutions at different concentrations (0, 1.5, 2.0, or

2.5%) and at different temperatures (20, 40, or 50ºC) for 3 min (Dikici, Koluman, and Calicioglu 2015), showed that decontamination of spinach with lactic acid solutions prepared at increased mild temperatures (40 and 50ºC) might be an effective safety measure in preventing public health risks associated with these products contaminated with Shiga toxin-producing *Escherichia coli*. Other evaluated combinations of treatments against *E. Coli, Salmonella Typhimurium* and were treatment of chlorine dioxide gas and aerosolized peracetic acid, for inactivation on spinach leaves (Park and Kang 2015) combined with UV-C and ClO_2 gas (Park, Kang, and Kang 2018). In the first study, the combined treatment of ClO_2 gas and aerosolized PAA produced injured cells of three pathogens on spinach leaves.

For UV-C and ClO_2 combined treatments, spinach samples were subjected to simultaneous treatment of UV-C and ClO_2 gas (5 or 10 ppm) for 5, 10, 15, or 20 min. All experiments were performed at 22 °C, and RH of the treatment chamber was adjusted with distilled water to 90% during treatment. This experiment resulted in additive effects: antimicrobial effects of UVC-ClO_2 gas (5 ppm) treatment were not superior to those of individual treatments during 15 min. The total microbial inactivation of the combined treatment was not significantly different from the sum of individual treatments and there was no significant impact on quality parameters.

Recently, Mukhopadhyay et al. (2019) studied the effects of pulsed light (PL) and sanitizer wash combination on inactivation of *Escherichia coli*, microbial loads and apparent quality of spinach leaves. A new formula of sanitizer (HEN) consisting of hydrogen peroxide, EDTA and Nisin were tested. Spinach leaves were spot inoculated on surface before treating with PL (1–63 J cm^{-2}), HEN sanitizer wash (2min) or their combinations. PL inactivation was influenced significantly at low doses. Treatment dose of 15.75 J cm^{-2}, equivalent to 15 s intense PL treatment, was found optimal, above which adverse quality effects were evident.

The widest combination of postharvest technologies applied to ready-to-eat spinach was reported (Ngnitcho et al. 2017). These authors studied the effects of chemical treatment using slightly acidic electrolyzed water

(SAEW), fumaric acid (FA), or calcium oxide (CaO) and physical treatment using ultrasonication (US), micro-bubbles (MB), or ultraviolet (UV) to inactivate bacterial pathogens *Listeria monocytogenes*, *Escherichia coli* O157:H7, *Staphylococcus aureus*, and *Salmonella* spp. on spinach. Spinach leaves were inoculated with bacterial pathogens (~9 log cfu mL^{-1}) and immersed in distilled water (DW), SAEW (30 ppm), FA (0.5%), or CaO (0.2%) alone or in combination, at 23 °C for 3 min, followed by treatment with US, MB for 3 min, or UV for 10 min. A combination of CaO, SAEW, and FA sanitizers appears to be a favorable method to ensure microbial safety as well as to prolong the shelf-life of fresh produce with little damaging effect on their quality. Using physical treatment such as ultrasonication can enhance microbial reduction but at the same time, it might impair the quality of produce compared to other treatments. Alternatively, CaO +FA + SAEW without physical treatment could be used to increase microbial reduction and improve fresh produce quality.

Relative Humidity and Temperature during Processing

Temperature and relative humidity (RH) are two conditions that influence the vapour pressure deficit leading to water loss from spinach leaves to the surrounding ambient air. Optimal conditions of temperature and RH after harvest and along processing, are fundamental to keep the quality and extend shelf life but not always possible to achieve (Ghidelli and Pérez-Gago 2018). Baby leaves with a reduced water content before processing could maintain quality better because they are less susceptible to damage during processing and during storage, when compared to leaves with higher moisture content (Oliveira et al. 2016).

One of the common techniques applied to remove excess water from leaves surface is trough hot air drying. However, this can lead to problems such as deterioration in the taste, colour and nutritional content and loss of solutes to the surface, due to the long drying period and high temperature used (Onwude et al. 2016). One of the alternatives to the need of drying or having a product with lower superficial water content is through the manipulation of variables such as temperature and RH. For the optimization of this manipulation, a predictive tool through mathematical modeling is fundamental, where industrial processing conditions can be simulated to control several dependent variables like microbial spoilage (Putnik et al. 2017), drying kinetics (Onwude et al. 2016), temperature effect on ready-to-eat leafy salads (Tsironi et al. 2017; Song, Liu, and Jaganathan 2016).

Besides being two of the most studied variables that impact ready-to-eat vegetables, there is a lack of information about the impact of temperature/RH during the waiting period between washing/drying and product packaging. A recent study was performed with baby spinach leaves exposed to different RH levels (55, 75 and 95%), temperatures (3, 5 and 7 °C) and waiting times after processing (1, 2.5 and 4 h). For Mediterranean summer spinach production, the industrial conditions of 4 °C, 83% RH and 4 h would provide leaves with higher quality before packaging, with a lower degree of surface wetness and would minimize further condensation inside the bags during distribution and retailing (Oliveira et al. 2016).

The importance of storage conditions on spinach quality leads to the development of new tools like hyperspectral imaging with chemometrics. This will allow providing an alternative for real-world on-line vegetables quality monitoring. The visualization of the distribution map of pigments content (chlorophylls and carotenoids) on spinach leaves stored at 4ºC for 9 days and 20ºC for 2 days, were acquired by applying the partial least squares (PLS) models on pixels within the hyperspectral image (C. Zhang et al. 2017).

PACKAGING CONDITIONS

Modified Atmosphere Packaging

Packaging is the final operation of processing of ready-to-eat products, that allows the products to be protected and safely distributed to consumers

(Wilson et al. 2019). The quality characteristics and shelf life extension are highly affected by packaging conditions and material, which will determine atmosphere, humidity, oxidation reactions and flavor diffusion into the product (Sartori and Menegalli 2016). Controlled atmosphere packaging (CAP) and modified atmosphere packaging (MAP) for ready-to-eat products are novel technologies that offer advantages to preserve the freshness, quality and shelf-life extension during storage (Boyacioglu et al. 2016).

MAP can be passive where the fresh product is packed with atmospheric pressure and the product naturally modifies the surrounding atmosphere by consuming O_2 and producing CO_2 (Saini, Ko, and Keum 2017). In active MAP the desired atmosphere conditions are created in headspace before sealing, but the final atmosphere will eventually be a function of the same factors that in passive MAP (Mahajan et al. 2017).

Recently, mathematical models have been proposed for predicting how packaging variables, such as gas composition, humidity and water vapor condensation dynamics, affect product quality under different environmental conditions that mimic supply chain (Jalali et al. 2017). The use of atmosphere with reduced O_2 concentration (1–5%) and higher CO_2 concentration (5–10%) was proposed as beneficial for decreasing microbial growth and retarding host tissue advanced stages of senescence, making them more resistant to microorganisms attack (Mahajan et al. 2017).

One of the problems associated with packaging baby spinach under MAP with low O_2 and high CO_2 concentrations is the development of off-odor compounds, without a direct effect in the visual aspect of the product. Baby leaf spinach packaged in polypropylene film with permeability to O_2 and CO_2 and stored 14 days at 7°C and 97% RH, allowed the creation of a passive modified atmosphere. At the end of 14 days, O_2 decreased to 0.3 kPa and CO_2 increase to 9.3 kPa promoting the development of off-odors compounds. Lipid peroxidation evidence reveal the presence of alcohols, sulfur compounds from amino acid degradation, and alkanes from lipid autoxidation processes or carotenoid degradation. Sensory panel described biomarkers of off-odors like rotten (alcohols 1-pentanol, (Z)-3-hexen-1-ol and 1-octen-3-ol), fishy notes (dimethyl sulfide) and sulfur compounds (Díaz-Mula et al. 2017).

Another study described the impact of MAP (5% O_2; 15% CO_2; balance N_2) on quality of baby spinach leaves when stored at different temperatures (4, 10 and 20ºC) during 12 days, by opposition to a control (78% N_2; 21% O_2). Both atmosphere conditions in package headspace changed faster as temperature rised, but a rapid decrease in O_2 and increase in CO_2 was observed for storage at 20ºC, under atmospheric conditions. Quality parameters such as weight loss, sensory quality, minerals, flavonoids, antioxidant capacity and headspace gas, started to decline from day 3 of storage until reaching unacceptable marketability after 6 days of storage under 20ºC under 21% O_2. However, spinach samples stored at 5% O_2, under cold conditions, still had overall quality for acceptable marketability by the end of 6 days. The ideal conditions found for baby spinach leaves commercialization were under CA (5% O_2; 15% CO_2; balance N_2) at 4ºC for 9 d (Mudau et al. 2018).

Nevertheless, nowadays MAP design is considered as the most crucial factor for deciding target gas barrier properties aiming to achieve an equilibrium in modified atmosphere through the product respiration rate. However, it is also essential to take into consideration other factors such as humidity rate inside the package, in order to avoid condensation and/or mold and bacterial development (Mahajan et al. 2017).

Packaging Films and Edible Coatings

Commercially, processed ready-to-eat leafy vegetables are currently packaged in perforated film bags or trays to allow adequate respiration rate and guarantee product quality and safety. The most common material used in ready-to-eat leafy vegetables packaging is plastic due to their relatively low price, mechanical resistance, heat sealability, shape versatility and rigidity (Otoni et al. 2017). However, plastic is non-renewable and mostly non-biodegradable. In this context, there has been worldwide demand for replacing plastics for a renewable and biodegradable polymer. Edible films and coatings can be originated from polysaccharides, polypeptides, lipids or mixtures of these biopolymers (Sartori and Menegalli 2016). They are renewable and biodegradable and have excellent film-forming properties and, most of them are edible therefore, can be used as matrices for edible films and coatings, which are supposed to be ingested with the food. Among polysaccharides, starch, chitosan, cellulose, polycaprolactone and polylactic acid have received particular attention because their abundant and renewable or with biodegradable properties (Abdul Khalil et al. 2018).

Therefore, in last years, edible films and coatings with bioactive molecules incorporated for packaging systems have gained a lot of interest in the food packaging industry, as eco-friendly alternatives to conventional plastic packaging and approach to extend the shelf-life and quality of ready-to-eat products (Issa, Ibrahim, and Tahergorabi 2017). Edible coatings can provide a semipermeable barrier to gases and water vapour, which might translate in a reduction in respiration rate, enzymatic browning, and water loss. Their protective function may also be enhanced with the addition of ingredients, such as antioxidants, antimicrobials, flavors, beyond others (Ghidelli and Pérez-Gago 2018). Films and coatings may effectively carry food additives, including antioxidants, vitamins, colourants, flavors, and antimicrobial agents, thereby enhancing the integrity of the food they contain (Sartori and Menegalli 2016).

Antimicrobial edible films were developed by applying a gelatin extracted from olive flounder bones (OBG), which is a fish-processing by-product, and zinc oxide nanoparticles (ZnO). Spinach leaves packaged with OBG film containing 3% of Zno revealed a decrease in *Listeria monocytogenes* growth during one week of storage at 4 °C and 60% RH. Vitamin C content decreased for all packaged samples during storage, but the lower decrease was reported for OBG packages. This was probably due to the light barrier property of the OBG-ZnO film. The chlorophyll content did not change significantly for all packaged samples (Beak, Kim, and Song 2017).

A combination of sweet potato starch (SPS), montmorillonite nanoclay (MMT) and thyme essential oil (TEO) was used to develop an active, antibacterial and biodegradable film for baby spinach leaves packaging. As

SPS does not have the mechanical and barrier properties matching those plastics, nano-scale particles of MMT were added to improve this property. Results showed that TEO (2, 4 or 6%) incorporation in the film with SPS-MTT significantly reduced the population of *Escherichia coli* and *Salmonella typhimurium* on fresh baby spinach leaves to non-detectable levels within 8 days, at 4ºC. While control samples (without essential oil) had an increase of 2 log colony forming unit g^{-1}. The sensory analysis scores for odor, color and appearance for spinach wrapped in a film containing TEO were higher than those unwrapped, which was probably due to the aromatic effect of thyme essential oil. This study thus suggests that use of essential oils with antibacterial effect has the potential for direct incorporation into the SPS film, developing antimicrobial nanocomposite films for food packaging applications (Issa, Ibrahim, and Tahergorabi 2017).

Edible films or coatings are very promising systems for packaging fresh spinach leaves, improving their quality and shelf life during processes and storage, being a renewable and biodegradable replacement to plastic in packaging.

STORAGE CONDITIONS AFTER PACKAGING

Storage Temperature, Time and Relative Humidity

The component of the postharvest environment that has the major impact on the visual, compositional, and eating quality of spinach is the binomious temperature/time. A good management of this factor will allow delaying the product deterioration. In addition, this can be achieved when the product is cooled to its optimal temperature immediately after harvesting. Generally, lower temperatures are associated with longer storage life, by retarding ageing, softening, textural and colour changes, as well as slowing undesirable metabolic changes, moisture loss, and losses due to pathogen invasion (Brasil and Siddiqui 2018).

Temperature is one of the main critical factors for transpiration, since a small temperature fluctuation might influence the formation of water droplets underneath the film packaging leading to deterioration of packaged product (Volpe et al. 2018).

Packaged baby spinach stored at 4 and 22ºC, for 12 days, revealed that refrigerated conditions allowed keep higher quality until 6 days of storage, while at 22ºC nutritional parameters declined after 2 days storage. Refrigerated samples revealed higher amount in minerals magnesium (Mg), zinc (Zn), and iron (Fe), higher stability in carotenoids, total phenolic compounds and vitamin C, which reflects in the antioxidant capacity. This study demonstrated that quality of baby spinach deteriorates as storage time and temperature increase (Mudau et al. 2015).

In general, it is also known that ready-to-eat fruit and vegetable products should be stored in high relative humidity conditions, so the moisture loss will be minimized, but it cannot be so high that it creates condensation that might increase microbiological development (Brasil and Siddiqui 2018). Readyto-eat vegetables are usually stored in polypropylene bags at recommended storage conditions of 85-98% RH (Bovi et al. 2016).

The relative humidity inside package is directly related with water loss from the fresh produce, since higher relative humidity around the leaves will reduce the effect of water vapour pressure deficit. However, at near-saturation conditions minor fluctuation in temperature could lead to condensation inside the packaging resulting in accelerated microbial growth, sliminess, decay, and induced physiological disorders (Caleb et al. 2016).

Moisture accumulation in the package involves water transfer from inter and intracellular space to the skin of the product, water evaporation from the outer surface layer and convective mass transfer of the moisture to the surroundings. Under saturated RH condition, transpiration is caused by internal heat generation due to respiration (Volpe et al. 2018). Understanding the physiological response of individual fresh horticultural produce towards optimum packaging/storage system design with adequate humidity control is one of the keys to achieving the postharvest technology goals (Bovi et al. 2016).

After packaged, leafy vegetables like spinach are stored in factories before being dispatched to the supermarket. This period can be detrimental for a good quality product reaching the supermarket. A recent study was performed with packaged baby spinach leaves which were stored at different RH levels (55, 75 and 95%), temperatures (3, 5 and 7 °C) and storage times of 24, 48 and 72 h. This study reveals that conditions of 3 °C, 74% RH for a maximum of 1 day before dispatch, allowed getting lower moisture content, lower electrolyte leakage with higher content of chlorophyll and oxygen concentration inside the bags (Oliveira et al. 2016). This study demonstrated the potential of mathematical models for predicting storage conditions.

Light Conditions

Fresh-cut leafy vegetables are exposed to various light conditions during shelf life, and the effects of light exposure on quality changes during storage have been reported for a number of crops (Saini, Ko, and Keum 2017). The intensity of light which leafy vegetables are exposed can induce stress, leading to the production of oxygen reactive species (ROS) that will destroy important antioxidants such as ascorbic acid, carotenoids and phenolic compounds. On the other hand, the dose of stress (as intensity of light) can improve the nutritional quality of leafy vegetables, by stimulating the antioxidant response without causing tissue damage and accelerate leaf senescence.

Light exposure has been reported to reduce the solute leakage in spinach leaves when compared with storage in the dark, suggesting improved leaf resilience. Continuous low-intensity light (20 – 27 μmol m^{-2} s^{-1}) during spinach storage reduced ascorbic acid loss, probably due to higher availability of ascorbic acid precursors. The nutritional quality and texture of spinach leaves stored under continuous (24 h) low-intensity light at 4ºC was found to be improved compared with dark-stored leaves, whereas enhanced degradation of carotenoids and chlorophylls has been observed. High-intensity light exposure resulted in oxidative stress, causing tissue damage and quality loss as evidenced by increased membrane damage and water loss (Glowacz et al. 2015).

Light exposure can retard tissue browning, delayed yellowing and have a positive effect on nutritional quality. In spinach leaves, the endogenous pool of some vitamins (such as ascorbic acid, folate) showed higher values when leaves where stored under visible light than in the dark. Light could enhance ascorbic acid synthesis due to the increase of the photosynthetic capacity that induce the availability of soluble carbohydrates, especially glucose, enabling them to contribute to the control of the ascorbic acid pool size (Charles et al. 2018).

The environment in which leafy vegetables are confined can be important to the final content of phytochemicals, since these continue active after harvesting. The carotenoids (violaxanthin, antheraxanthin and zeaxanthin) are compounds responsible for plants photoprotection and they regulate thermal dissipation of excess light energy. A recent study revealed a daily course of spinach pigment change under supermarket conditions (1 h before shelving lights switched on and two hours before shelving lights switched off; $\approx 16 \pm 2.6$ µmol photons $m^{-2}s^{-1)}$. The chlorophylls a and b showed a significant decrease during the day, while total carotenoids showed no significant changes during the day (Esteban et al. 2019). A better understanding of light effect in spinach quality during storage could allow better quality management within the supply chain.

Conclusion

Achieving maximum postharvest quality goes through understanding, predicting and managing various postharvest factors impacting product quality. There are many different technologies to reduce quality loss and increase safety of processed spinach. Levels of biologically active compounds and other quality characteristics like appearance are affected by most processing and storage factors. Harvested spinaches transported from farm to the final consumer have to be continuously in controlled conditions once the distribution chain involves many challenges and the product continues both metabolic and physiological activities after harvest. Many recent studies have focus on recent technologies for decontamination but attention must be paid on the effect of spinach bioactive compounds and consequently on nutritional quality. Thus, strict control of optimum conditions during cold chain and storage, with MAP as supplement, are decisive factors for maintaining quality. Further research is necessary to understand the condensation dynamics and develop moisture removal technologies in packaged products. The equilibrium within several postharvest disinfection technologies and respective combination with/of factors such as temperature, relative humidity, modified atmosphere and packaging film have been the pipeline for keeping postharvest quality of ready-to-eat spinach.

References

Abdul Khalil, H P S, Aparajita Banerjee, Chaturbhuj K Saurabh, Y Y Tye, A B Suriani, A Mohamed, A A Karim, Samsul Rizal, and M T Paridah. 2018. "Biodegradable Films for Fruits and Vegetables Packaging Application: Preparation and Properties." *Food Engineering Reviews* 10 (3): 139–53. https://doi.org/10.1007/s12393-018-9180-3.

Agüero, M Victoria, Rosa J Jagus, Olga Martín-Belloso, and Robert Soliva-Fortuny. 2016. "Surface Decontamination of Spinach by Intense Pulsed Light Treatments: Impact on Quality Attributes." *Postharvest Biology and Technology* 121: 118–25. https://doi.org/ https://doi.org/10.1016/j.postharvbio.2016.07.018.

Ariffin, Siti H, Kostas Gkatzionis, and Serafim Bakalis. 2017. "Leaf Injury and Its Effect towards Shelf-Life and Quality of Ready-to-Eat (RTE) Spinach." *Energy Procedia* 123: 105–12. https://doi.org/10.1016/ j.egypro.2017.07.265.

Beak, Songee, Hyeri Kim, and Kyung Bin Song. 2017. "Characterization of an Olive Flounder Bone Gelatin-Zinc Oxide Nanocomposite Film and Evaluation of Its Potential Application in Spinach Packaging."

Journal of Food Science 82 (11): 2643–49. https://doi.org/10.1111/1750-3841.13949.

Bovi, Graziele G, Oluwafemi J Caleb, Manfred Linke, Cornelia Rauh, and Pramod V Mahajan. 2016. "Transpiration and Moisture Evolution in Packaged Fresh Horticultural Produce and the Role of Integrated Mathematical Models: A Review." *Biosystems Engineering* 150: 24–39. https://doi.org/https://doi.org/10.1016/j.biosystemseng.2016.07.013.

Boyacioglu, O, A Sulakvelidze, M Sharma, and I Goktepe. 2016. "Effect of a Bacteriophage Cocktail in Combination with Modified Atmosphere Packaging in Controlling Listeria Monocytogenes on Fresh-Cut Spinach." *Irish Journal of Agricultural and Food Research* 55 (1): 74–79. https://doi.org/https://doi.org/10.1515/ijafr-2016-0007.

Brasil, Isabella M, and Mohammed Wasim Siddiqui. 2018. "Chapter 1 - Postharvest Quality of Fruits and Vegetables: An Overview." In, edited by Mohammed Wasim B T - Preharvest Modulation of Postharvest Fruit and Vegetable Quality Siddiqui, 1–40. Academic Press. https://doi.org/https://doi.org/10.1016/B978-0-12-809807-3.00001-9.

Caleb, Oluwafemi J, Kathrin Ilte, Antje Fröhling, Martin Geyer, and Pramod V Mahajan. 2016. "Integrated Modified Atmosphere and Humidity Package Design for Minimally Processed Broccoli (Brassica Oleracea L. Var. Italica)." *Postharvest Biology and Technology* 121: 87–100. https://doi.org/https://doi.org/10.1016/j.postharvbio.2016.07.016.

Cefola, Maria, and Bernardo Pace. 2015. "Application of Oxalic Acid to Preserve the Overall Quality of Rocket and Baby Spinach Leaves during Storage." *Journal of Food Processing and Preservation* 39 (6): 2523–32. https://doi.org/10.1111/jfpp.12502.

Charles, Florence, Phrutiya Nilpraprruck, David Roux, and Huguette Sallanon. 2018. "Visible Light as a New Tool to Maintain Fresh-Cut Lettuce Post-Harvest Quality." *Postharvest Biology and Technology* 135: 51–56. https://doi.org/https://doi.org/10.1016/j.postharvbio.2017.08.024.

Dewhirst, Rebecca A, Graham J J Clarkson, Steve D Rothwell, and Stephen C Fry. 2017. "Novel Insights into Ascorbate Retention and Degradation during the Washing and Post-Harvest Storage of Spinach and Other Salad Leaves." *Food Chemistry* 233: 237–46. https://doi.org/10.1016/j.foodchem.2017.04.082.

Díaz-Mula, Huertas M, Alicia Marín, María José Jordán, and María I Gil. 2017. "Off-Odor Compounds Responsible for Quality Loss of Minimally Processed Baby Spinach Stored under MA of Low O2 and High CO2 Using GC–MS and Olfactometry Techniques." *Postharvest Biology and Technology* 129: 129–35. https://doi.org/10.1016/j.postharvbio.2017.03.018.

Dikici, Abdullah, Ahmet Koluman, and Mehmet Calicioglu. 2015. "Comparison of Effects of Mild Heat Combined with Lactic Acid on Shiga Toxin Producing Escherichia Coli O157:H7, O103, O111, O145 and O26 Inoculated to Spinach and Soybean Sprout." *Food Control* 50: 184–89. https://doi.org/10.1016/j.foodcont.2014.08.038.

Esteban, Raquel, Javier Buezo, José M Becerril, and José I García-Plazaola. 2019. "Modified Atmosphere Packaging and Dark/Light Refrigerated Storage in Green Leafy Vegetables Have an Impact on Nutritional Value." *Plant Foods for Human Nutrition.* https://doi.org/10.1007/s11130-018-0705-y.

Finten, G, M V Agüero, and R J Jagus. 2017. "Citric Acid as Alternative to Sodium Hypochlorite for Washing and Disinfection of Experimentally-Infected Spinach Leaves." *LWT - Food Science and Technology* 82: 318–25. https://doi.org/10.1016/j.lwt.2017.04.047.

Ghidelli, Christian, and María B Pérez-Gago. 2018. "Recent Advances in Modified Atmosphere Packaging and Edible Coatings to Maintain Quality of Fresh-Cut Fruits and Vegetables." *Critical Reviews in Food Science and Nutrition* 58 (4): 662–79. https://doi.org/10.1080/10408398.2016.1211087.

Glowacz, Marcin, Lars M Mogren, John P H Reade, Andrew H Cobb, and James M Monaghan. 2015. "High- but Not Low-Intensity Light Leads to Oxidative Stress and Quality Loss of Cold-Stored Baby Leaf

Spinach." *Journal of the Science of Food and Agriculture* 95 (9): 1821–29. https://doi.org/10.1002/jsfa.6880.

Gu, Ganyu, Andrea Ottesen, Samantha Bolten, Padmini Ramachandran, Elizabeth Reed, Steve Rideout, Yaguang Luo, Jitendra Patel, Eric Brown, and Xiangwu Nou. 2018. "Shifts in Spinach Microbial Communities after Chlorine Washing and Storage at Compliant and Abusive Temperatures." *Food Microbiology* 73: 73–84. https://doi.org/10.1016/j.fm.2018.01.002.

Hussain, Peerzada R, Prashant Suradkar, Sadiya Javaid, Heena Akram, and Sadaf Parvez. 2016. "Influence of Postharvest Gamma Irradiation Treatment on the Content of Bioactive Compounds and Antioxidant Activity of Fenugreek (Trigonella Foenum–graceum L.) and Spinach (Spinacia Oleracea L.) Leaves." *Innovative Food Science & Emerging Technologies* 33: 268–81. https://doi.org/10.1016/j.ifset.2015.11.017.

Issa, Aseel, Salam A Ibrahim, and Reza Tahergorabi. 2017. "Impact of Sweet Potato Starch-Based Nanocomposite Films Activated With Thyme Essential Oil on the Shelf-Life of Baby Spinach Leaves." *Foods*. https://doi.org/10.3390/foods6060043.

Jalali, Ali, Sadegh Seiiedlou, Manfred Linke, and Pramod Mahajan. 2017. "A Comprehensive Simulation Program for Modified Atmosphere and Humidity Packaging (MAHP) of Fresh Fruits and Vegetables." *Journal of Food Engineering* 206: 88–97. https://doi.org/10.1016/j.jfoodeng.2017.03.007.

Jiang, Yunbin, Kimberly Sokorai, Georgios Pyrgiotakis, Philip Demokritou, Xihong Li, Sudarsan Mukhopadhyay, Tony Jin, and Xuetong Fan. 2017. "Cold Plasma-Activated Hydrogen Peroxide Aerosol Inactivates Escherichia Coli O157:H7, Salmonella Typhimurium, and Listeria Innocua and Maintains Quality of Grape Tomato, Spinach and Cantaloupe." *International Journal of Food Microbiology* 249: 53–60. https://doi.org/10.1016/j.ijfoodmicro.2017.03.004.

Kang, Ji-Hoon, Jun-Beom Park, and Kyung Bin Song. 2019. "Inhibitory Activities of Quaternary Ammonium Surfactants against Escherichia Coli O157:H7, Salmonella Typhimurium, and Listeria Monocytogenes

Inoculated on Spinach Leaves." *LWT* 102: 284–90. https://doi.org/10.1016/j.lwt.2018.12.046.

Karaca, Hakan, and Y Sedat Velioglu. 2014. "Effects of Ozone Treatments on Microbial Quality and Some Chemical Properties of Lettuce, Spinach, and Parsley." *Postharvest Biology and Technology* 88: 46–53. https://doi.org/10.1016/j.postharvbio.2013.09.003.

Liao, Chunli, Xiaobo Liu, Along Gao, Anfang Zhao, Jiyong Hu, and Bingbing Li. 2016. "Maintaining Postharvest Qualities of Three Leaf Vegetables to Enhance Their Shelf Lives by Multiple Ultraviolet-C Treatment." *LWT* 73: 1–5. https://doi.org/10.1016/j.lwt.2016.05.029.

Mahajan, Pramod V, Oluwafemi J Caleb, Maria I Gil, Hidemi Izumi, Giancarlo Colelli, Christopher B Watkins, and Manuela Zude. 2017. "Quality and Safety of Fresh Horticultural Commodities: Recent Advances and Future Perspectives." *Food Packaging and Shelf Life* 14: 2–11. https://doi.org/10.1016/j.fpsl.2017.08.001.

Meireles, Ana, Efstathios Giaouris, and Manuel Simões. 2016. "Alternative Disinfection Methods to Chlorine for Use in the Fresh-Cut Industry." *Food Research International* 82: 71–85. https://doi.org/10.1016/j.foodres.2016.01.021.

Mudau, Ambani R, Mpumelelo M Nkomo, Puffy Soundy, Hintsa T Araya, Wonder Ngezimana, and Fhatuwani N Mudau. 2015. "Influence of Postharvest Storage Temperature and Duration on Quality of Baby Spinach." *HortTechnology Hortte* 25 (5): 665–70. https://doi.org/10.21273/HORTTECH.25.5.665.

Mudau, Ambani R, Puffy Soundy, Hintsa T Araya, and Fhatuwani N Mudau. 2018. "Influence of Modified Atmosphere Packaging on Postharvest Quality of Baby Spinach (Spinacia Oleracea L.) Leaves." *HortScience Horts* 53 (2): 224–30. https://doi.org/10.21273/HORTSCI12589-17.

Mukhopadhyay, Sudarsan, Kimberly Sokorai, Dike O Ukuku, Xuetong Fan, Modesto Olanya, and Vijay Juneja. 2019. "Effects of Pulsed Light and Sanitizer Wash Combination on Inactivation of Escherichia Coli O157:H7, Microbial Loads and Apparent Quality of Spinach Leaves."

Food Microbiology 82: 127–34. https://doi.org/10.1016/j.fm.2019.01.022.

Murray, Kayla, Fan Wu, John Shi, Sophia Jun Xue, and Keith Warriner. 2017. "Challenges in the Microbiological Food Safety of Fresh Produce: Limitations of Post-Harvest Washing and the Need for Alternative Interventions." *Food Quality and Safety* 1 (4): 289–301. https://doi.org/10.1093/fqsafe/fyx027.

Ngnitcho, Paul-François Kounkeu, Imran Khan, Charles Nkufi Tango, Mohammad Shakhawat Hussain, and Deog Hwan Oh. 2017. "Inactivation of Bacterial Pathogens on Lettuce, Sprouts, and Spinach Using Hurdle Technology." *Innovative Food Science & Emerging Technologies* 43: 68–76. https://doi.org/10.1016/j.ifset.2017.07.033.

Oliveira, Ana, Pedro M Castro, Ana L Amaro, Jason de Sain, and Manuela Pintado. 2016. "Optimization of Temperature, Relative Humidity and Storage Time before and after Packaging of Baby Spinach Leaves Using Response Surface Methodology." *Food and Bioprocess Technology* 9 (12): 2070–79. https://doi.org/10.1007/s11947-016-1785-z.

Otoni, Caio G, Roberto J Avena-Bustillos, Henriette M C Azeredo, Marcos V Lorevice, Márcia R Moura, Luiz H C Mattoso, and Tara H McHugh. 2017. "Recent Advances on Edible Films Based on Fruits and Vegetables—A Review." *Comprehensive Reviews in Food Science and Food Safety* 16 (5): 1151–69. https://doi.org/10.1111/1541-4337.12281.

Papachristodoulou, Michalis, Athanasios Koukounaras, Anastasios S Siomos, Aglaia Liakou, and Dimitrios Gerasopoulos. 2018. "The Effects of Ozonated Water on the Microbial Counts and the Shelf Life Attributes of Fresh-Cut Spinach." *Journal of Food Processing and Preservation* 42 (1): e13404. https://doi.org/10.1111/jfpp.13404.

Park, Sang-Hyun, and Dong-Hyun Kang. 2015. "Combination Treatment of Chlorine Dioxide Gas and Aerosolized Sanitizer for Inactivating Foodborne Pathogens on Spinach Leaves and Tomatoes." *International Journal of Food Microbiology* 207: 103–8. https://doi.org/10.1016/j.ijfoodmicro.2015.04.044.

———. 2017. "Influence of Surface Properties of Produce and Food Contact Surfaces on the Efficacy of Chlorine Dioxide Gas for the Inactivation of Foodborne Pathogens." *Food Control* 81: 88–95. https://doi.org/10.1016/j.foodcont.2017.05.015.

Park, Sang-Hyun, Jun-Won Kang, and Dong-Hyun Kang. 2018. "Inactivation of Foodborne Pathogens on Fresh Produce by Combined Treatment with UV-C Radiation and Chlorine Dioxide Gas, and Mechanisms of Synergistic Inactivation." *Food Control* 92: 331–40. https://doi.org/10.1016/j.foodcont.2018.04.059.

Pasquali, Frederique, Alexandros Ch. Stratakos, Anastasios Koidis, Annachiara Berardinelli, Chiara Cevoli, Luigi Ragni, Rocco Mancusi, Gerardo Manfreda, and Marcello Trevisani. 2016. "Atmospheric Cold Plasma Process for Vegetable Leaf Decontamination: A Feasibility Study on Radicchio (Red Chicory, Cichorium Intybus L.)." *Food Control* 60: 552–59. https://doi.org/10.1016/j.foodcont.2015.08.043.

Roberts, Joseph L, and Régis Moreau. 2016. "Functional Properties of Spinach (Spinacia Oleracea L.) Phytochemicals and Bioactives." *Food Funct.* 7 (8): 3337–53. https://doi.org/10.1039/C6FO00051G.

Saini, Ramesh Kumar, Eun Young Ko, and Young-Soo Keum. 2017. "Minimally Processed Ready-to-Eat Baby-Leaf Vegetables: Production, Processing, Storage, Microbial Safety, and Nutritional Potential." *Food Reviews International* 33 (6): 644–63. https://doi.org/10.1080/87559129.2016.1204614.

Saini, Ramesh Kumar, Xiao Min Shang, Eun Young Ko, Jeong Hee Choi, and Young-Soo Keum. 2016. "Stability of Carotenoids and Tocopherols in Ready-to-Eat Baby-Leaf Lettuce and Salad Rocket during Low-Temperature Storage." *International Journal of Food Sciences and Nutrition* 67 (5): 489–95. https://doi.org/10.3109/09637486.2016.1172059.

Sartori, Tanara, and Florencia Cecilia Menegalli. 2016. "Development and Characterization of Unripe Banana Starch Films Incorporated with Solid Lipid Microparticles Containing Ascorbic Acid." *Food Hydrocolloids* 55: 210–19. https://doi.org/10.1016/j.foodhyd.2015.11.018.

Shynkark, Mykola V, Taras Pyatkovskyy, Hussein M Mohamed, Ahmed E Yousef, and Sudhir K Sastry. 2015. "Physics of Fresh Produce Safety: Role of Diffusion and Tissue Reaction in Sanitization of Leafy Green Vegetables with Liquid and Gaseous Ozone-Based Sanitizers." *Journal of Food Protection* 78 (12): 2108–16. https://doi.org/10.4315/0362-028X.JFP-15-290.

Shynkaryk, Mykola V, Taras I Pyatkovskyy, Ahmed E Yousef, and Sudhir K Sastry. 2016. "Gaseous Ozone Treatment of Baby Spinach within the Existing Production Chain for Inactivation of Escherichia Coli O157:H7." *Journal of Food Engineering* 191: 10–18. https://doi.org/10.1016/j.jfoodeng.2016.06.025.

Stratakos, Alexandros Ch., Mark Linton, Margaret F Patterson, and Anastasios Koidis. 2015. "Effect of High-Pressure Processing on the Shelf Life, Safety and Organoleptic Characteristics of Lasagne Ready Meals during Storage at Refrigeration and Abuse Temperature." *Innovative Food Science & Emerging Technologies* 30: 1–7. https://doi.org/10.1016/j.ifset.2015.05.010.

Stratakos, Alexandros Ch., Mark Linton, Girum Tadesse Tessema, Taran Skjerdal, Margaret F Patterson, and Anastasios Koidis. 2016. "Effect of High Pressure Processing in Combination with Weissella Viridescens as a Protective Culture against Listeria Monocytogenes in Ready-to-Eat Salads of Different PH." *Food Control* 61: 6–12. https://doi.org/10.1016/j.foodcont.2015.09.020.

Ufuk Kasım, M, and Rezzan Kasım. 2017. "Yellowing of Fresh-Cut Spinach (Spinacia Oleracea L.) Leaves Delayed by UV-B Applications." *Information Processing in Agriculture* 4 (3): 214–19. https://doi.org/10.1016/j.inpa.2017.05.006.

Volpe, Stefania, Pramod V Mahajan, Guido Rux, Silvana Cavella, and Elena Torrieri. 2018. "Condensation and Moisture Regulation in Packaged Fresh-Cut Iceberg Lettuce." *Journal of Food Engineering* 216: 132–37. https://doi.org/10.1016/j.jfoodeng.2017.08.015.

Wilson, Matthew Deas, Roger A Stanley, Alieta Eyles, and Tom Ross. 2019. "Innovative Processes and Technologies for Modified Atmosphere Packaging of Fresh and Fresh-Cut Fruits and Vegetables."

Critical Reviews in Food Science and Nutrition 59 (3): 411–22. https://doi.org/10.1080/10408398.2017.1375892.

Wyrwa, Joanna, and Anetta Barska. 2017. "Innovations in the Food Packaging Market: Active Packaging." *European Food Research and Technology* 243 (10): 1681–92. https://doi.org/10.1007/s00217-017-2878-2.

Yesil, Mustafa, David R Kasler, En Huang, and Ahmed E Yousef. 2017. "Efficacy of Gaseous Ozone Application during Vacuum Cooling against Escherichia Coli O157:H7 on Spinach Leaves as Influenced by Bacterium Population Size." *Journal of Food Protection* 80 (7): 1066–71. https://doi.org/10.4315/0362-028X.JFP-16-284.

Zhang, Chu, Qiaonan Wang, Fei Liu, Yong He, and Yuzhao Xiao. 2017. "Rapid and Non-Destructive Measurement of Spinach Pigments Content during Storage Using Hyperspectral Imaging with Chemometrics." *Measurement* 97: 149–55. ttps://doi.org/10.1016/j.measurement.2016.10.058.

Zhang, Min, Xiangyong Meng, Bhesh Bhandari, Zhongxiang Fang, and Huizhi Chen. 2015. "Recent Application of Modified Atmosphere Packaging (MAP) in Fresh and Fresh-Cut Foods." *Food Reviews International* 31 (2): 172–93. https://doi.org/10.1080/87559129.2014.981826.

In: Agricultural Research Updates. Vol. 28 ISBN: 978-1-53616-253-0
Editors: P. Gorawala, S. Mandhatri

Chapter 5

QUANTIFICATION OF NITRATE LEVELS IN SPINACH AND SIMUALTANEOUS DETERMINATION OF NITRITE

Teresa D'Amore, Aurelia Di Taranto, Giovanna Berardi, Valeria Vita and Marco Iammarino*

Department of Chemistry, Istituto Zooprofilattico Sperimentale della Puglia e della Basilicata, Foggia, Italy

ABSTRACT

Nitrates and nitrites are compounds that may accumulate in several foodstuffs, especially vegetables, due to their presence in the environment as consequence of the nitrogen cycle. Moreover, they are largely used in meat curing as food preservatives.

It is well known that nitrates and nitrites may exercise some harmful effects on humans. So, the levels of these compounds in foodstuffs have to be monitored by health authorities, continuously. These monitoring activities have been intensified on vegetables during the last years, since a significant increase of residual level of nitrate, especially in leafy

* Corresponding Author's E-mail: marco.iammarino@tin.it; marco.iammarino@izspb.it.

vegetables, has been verified. This increase seems especially due to the wide use of nitrogen fertilizers in agriculture.

The accumulation of nitrate in vegetables depends on N-fertilizer application timing, temperature, water relation, cultivar and other variables. Moreover, some studies have underlined the simultaneous presence of not negligible amounts of nitrite. This last finding is significant for food safety, since nitrite is involved in the synthesis of carcinogenic nitrosamines, directly.

The nitrite mechanism of synthesis in leafy vegetables is not clear. Presumably, it is a reduction product of endogenous nitrate.

In this chapter, the results obtained by analysing 30 samples of fresh, frozen and IV range spinach, with the aim to identify/quantify the levels of nitrate and nitrite, are described. The analyses were carried out by using a validated/accredited analytical method by ion chromatography with conductivity detection. The most important findings have been compared with other available data worldwide, and with nitrate/nitrite levels detected in other types of vegetables.

Very interesting results were obtained for both nitrate and nitrite. As expected, very high levels of nitrate were registered in spinach, where some concentrations higher than the European legal limit were verified. Concerning nitrite, detectable amounts of this contaminant were found, at levels up to 78.5 mg kg^{-1}.

This study confirmed that spinach is among the most nitrate-contaminated vegetables and that control is needed. Moreover, taking into account the nitrite levels detected in some samples, it is possible suggesting the introduction, by health and food safety authorities, of specific limits for this compound in leafy vegetables.

Keywords: nitrate, nitrite, ion chromatography, nitrate assimilation, monitoring studies

INTRODUCTION

Primary producers, especially plants, use Nitrogen for their nutrition. Dinitrogen N_2, that is the most abundant molecule in the atmosphere (78%), is converted into ammonia (Nitrogen fixation) to become available into soil and water. Then, in aerobic condition, it is converted into nitrite and nitrate (Ammonia oxidation* and Nitrite Oxidation**), during a global process known as N cycle. The success of this process depends on the activity of several microorganisms, such as oxidizers bacteria, archaea, and fungi [1, 2].

* $NH_3 + 1½O_2 \rightarrow NO_2^- + H_2O + H^+$
** $NO_2^- + ½O_2 \rightarrow NO_3^-$

Generally plant crops use a wide range of chemical N-fixed forms (organic and inorganic) for assimilating nitrogen from soil and water. Among these molecules, nitrate and nitrite are the most common naturally occurring forms. Nitrate is the most stable oxidized form, so it may be considered the major source of N for plants; instead nitrite is more chemically reactive and less distributed in nature. Recently, several researchers evaluated the amount of N fixed forms in plants, through N stable isotopes study, and have shown an increase. Other data are not in accordance, since they demonstrate that N concentration in plant tissues decreased globally in the last 30 years. Anyway, the concentration of nitrogen/nitrate in plants has been proposed as a quality indicator of environment state/pollution and human health [3].

Other nitrate and nitrite sources in the environment derive from inorganic and synthetic fertilizers, decaying plants, manure, transformation reactions of herbicides, bacterial activity, other living organisms and other anthropogenic activities [4]. Moreover, both NO_3^- and NO_2^-, particularly their potassium and sodium salts, are also used as food additives, due to their preservative, antimicrobial, antioxidant and flavouring functions, that ensure the quality of food, especially processed meat and dairy products [5].

In dietary intake, vegetables, water and cured meat are the main sources of nitrate and nitrite. It has been estimated that 80% of nitrate dietary exposure derives from vegetable consumption, instead the major source of nitrite are meat products [6]. In fact, in fresh vegetables, the concentration of nitrite is normally low, except for contaminated or bad stored products (incorrect temperature and moisture conditions). Mostly, it is possible to state that dark leafy vegetables contain a higher NO_3^- amount than other vegetables [7].

Among vegetables, there is high variability of Nitrogen-compounds concentration that depends on many factors such as type of plant and genotype, climatic condition, geographical region, humidity, soil, use of fertilizers, irrigation, temperature, water availability, light, pollution, planting density, maturity, harvesting time, storage time, carbon dioxide concentration and many others [8, 9].

Some of these have a marked influence on final nitrate/nitrite concentration, so we will get back on them during the chapter. For example, it has been observed that the nitrate amount decreases on boiling. This is probably due to its complete water solubility. For the same reason, also washing can lead to a modest reduction of nitrate level [10].

Regarding plant type, there are different classifications. The most complete and cited is the Santamaria's one, which compiled/drafted a classification of vegetables, based on nitrate content, summed up in Table 1 [11].

Table 1. Classification of vegetables based on nitrate content in mg kg^{-1} (fresh weight) (Santamaria et al. 2006)

Classification	Range of values (mg kg^{-1})	Vegetables varieties
VLC (Very Low Concentration)	< 200	artichoke, asparagus, broad bean, eggplant, garlic, onion, green bean, mushroom, pea, pepper, potato, summer squash, sweet potato, tomato, watermelon
LC (Low Concentration)	200 - 500	broccoli, carrot, cauliflower, cucumber, pumpkin, chicory
MC (Middle Concentration)	500 - 1000	cabbage, dill, turnip, savoy cabbage
HC (High Concentration)	1000 - 2500	celeriac, Chinese cabbage, endive, fennel, kohlrabi, leek, parsley
VHC (Very High Concentration)	> 2500	celery, cress, chervil, lettuce, red beetroot, spinach, rocket

It is easily noticeable that nitrate-accumulating vegetables belong to the families of Apiacee (celery, chervil, celeriac, fennel, parsley), Asteracee (lettuce, endive), Brassicaceae (cress, rocket, Chinese cabbage, kohlrabi) and Chenopodiaceae (red beetroot, spinach). Spinaches, object of our study, are among vegetables with very high nitrate content.

Moreover, as stated above, storage condition may have a significant influence on change of nitrate and nitrite content. In fact, spinach leaves, especially after storage period at room temperature, show a drop of nitrate amount and the increase of nitrite level. Cintya et al. evidenced that nitrite and nitrated increased of 43.86% and 16.48%, respectively, in spinaches stored for 48 hours at 25°C [12]. On the other hand, storage in refrigerator (until 7 days) does not cause substantial variation in nitrate/nitrite content. At low storage temperature, endogenous nitrate reductase enzymes are inactivated and bacterial activity is inhibited [13].

Anyway, several researchers assert that use and type of fertilizers (also high emission from agriculture), light period and geographical area (so, also the climate) are the leading factors, influence nitrate/nitrite content in all vegetables [14].

"Phytochemistry" of NO_3^- and NO_2^- in Spinach: Assimilation, Metabolism, Biochemical Functions and Regulation

The classification reported above may be considered only an overall indication but it is not sufficient because wide fluctuations in nitrogen molecules amount may occur, due to all reason already cited and for phylogenetic properties, phenotype and metabolic pathway.

Many researcher conducted on various leafy vegetables, particularly spinach, revealed that this variability is verifiable also in different parts of plants, because nitrate and nitrite concentration is directly related with "photosynthesis apparatus." In Figure 1, the spinach organs are shown and listed by decreasing nitrate amount [15]. In spinach the nitrate content in petioles may be more than double than the leaves. Some vegetables tend to accumulate nitrate only in early stage of plant life. On the contrary, the nitrate accumulation in spinach continues until maturity, and older leaves have the highest vacuolar nitrate concentration. Similarly, it was observed that in root tips and root cap nitrate concentration is higher than in primary roots [16, 17].

The NO_3^- uptake is mediated by several transporters in roots, then, it is transported and stored in other parts of the plant. In general, once absorbed by the roots, most of the nitrate is moved towards the leaves and the shoots [18]. The NO_3^- uptake systems were studied and divided into two principal classes: the high-affinity transport system, which can be constitutive (cHATS) or inducible (iHATS), and the low-affinity transport system (LATS), which correspond to two families of *Spinacia oleracea* nitrate transporters (*So*NRTs) NRT1 and NRT2, respectively. They are principally located on the plasma membrane and co-transport H^+ ions and NO_3^- into cytosol, causing a mild alkalinisation of medium (Stoichiometry 2:1). It was observed a negative feedback (negative regulation) of this uptake mediated by increased concentration in the medium of NH_4^+ and glutamine [19].

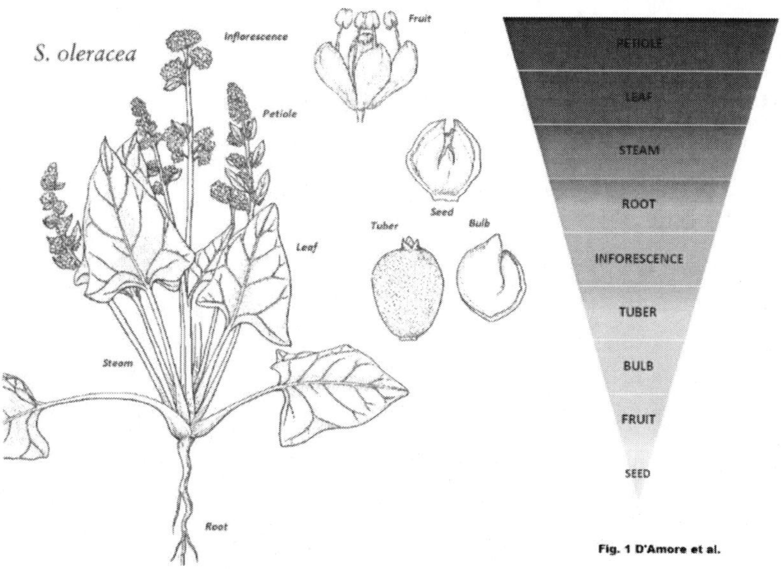

Figure 1. Vegetable-spinach organs, listed by decreasing nitrate amount.

Furthermore, the expression profiles of key genes, which encode nitrate transporters, were studied in spinach tissues. Using qRT-PCR it was shown that nitrate-rich soil induced the transcription of all *SoNRT1* and *SoNTR2* genes, not only in roots, but also in petioles and leaves (positive regulation). On the contrary, the excess of nitrate may inhibit its own uptake due to osmotic balance mechanisms [20].

Finally, cationic nutrients such as calcium, potassium, sodium and magnesium, also increase nitrate assimilation.

When absorbed by epidermal cells of root, nitrate can follow different roads. A part is immediately stored in vacuoles of roots, a part is moved through xylem tissues and in minimal part by phloem flow to other plant organs, and the residue is *in locum* reduced by Nitrate-reductase (NR) enzymes [21].

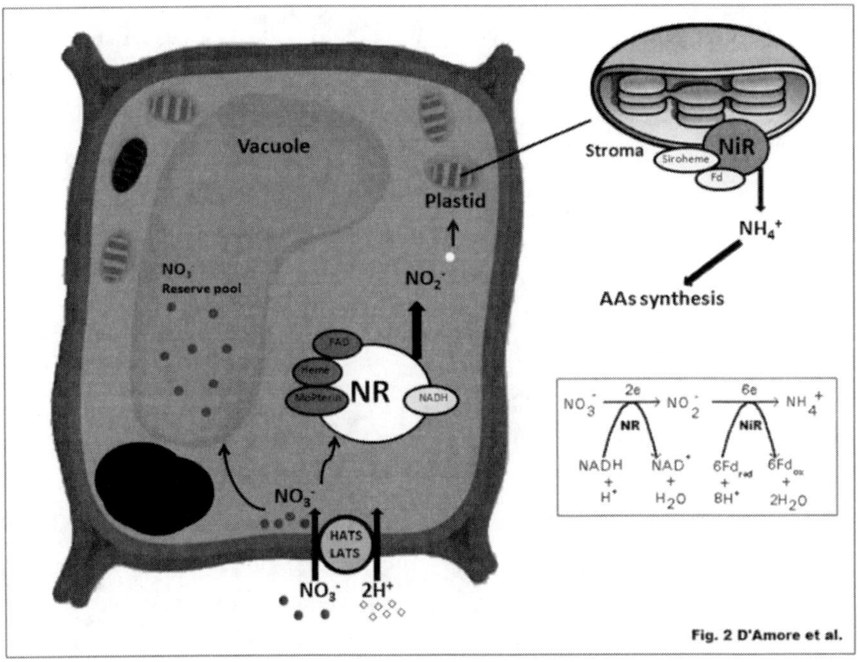

Figure 2. Plant cell: NO_3^-/NO_2^- uptake, assimilation, storage and metabolism. HATs = High Affinity Transport system; LATs = Low Affinity Transport system; NR = Nitrate reductase; NiR= Nitrite reductase; AAs = Amino acids.

Ott et al. have observed that most of nitrate is racked up in the mesophyll cells of spinach leaves, especially older ones, where the anion has four principal functions. In vacuoles, where it is stored up almost 80-90% of total nitrate present in a vegetal cell, it acts as storage compound (reserve pool), instead in cytosol has osmoregulation, signalling and metabolic functions. Surely, these observations about biochemical properties justify the different distribution of nitrate in plant parts [22].

When the plant is in its active period of growth (vegetative state), most of NO_3^- is bound by NR, located in cytosol and intracellular part of plasmatic membrane, and converted to NO_2^-. Thus, nitrite moves to the chloroplast, is further reduced to ammonium by nitrite-reductase (NiR), located in the stroma, and enters in the pathway of amino acid synthesis. In Figure 2 a schematic representation of NO_3^-/NO_2^- assimilation and functions is proposed. During this phase, it is possible finding slight high concentration of nitrite especially in petioles and leaves cells [23].

A key role is acted by NR, so the structure and function of this enzyme have been studied in deep in spinach plants. It is a dimero of about 900 amino acids, which uses FAD, MoPterin (or Molybdopterin) and a heme group as cofactors, and NADH as electron donor [24].

Some characteristics of two form of NR (one constitutive cNR and one inducible iNR), especially regarding their expression and activity regulation are interesting. In fact, we have previously explained that several factors influence the nitrate concentration in spinach, as confirmed by a lot of researches. Among these, light period exposition of spinach crops has a main role [25, 26].

These surveys are confirmed by molecular biology and electroforetical studies, which showed the effect of light as transcriptional inductor of NR genes (*NIA1* and *NIA2*), mediated by the phytocrome system. Moreover, NR has other mechanism of regulation, or rather phosphororylation and dephosphorylation events, mediated by kinase and phosphatase proteins. Enzymatic in vitro studies have highlighted that a specific amino acid residue Ser-543 can be dephosphorylated by phosphatase-2A (PP2A) in leaf-spinach NR, and that this event decreases NR activity. On the other hand, CDPK kinases active NR and this regulation are more effective with light [27].

Finally, one more post-traductional mechanism of regulation is discovered and consists in binding between phosforylated-NR and a protein of 14.3.3 family, which causes NR inactivation [28].

Instead NiR has a siroheme group and an iron-sulfur domain (4Fe-4S clusters), and uses Ferredoxin (Fd) as electron donor. Ferredoxin-dependent nitrite reductase has been isolated from spinach and characterized by X-ray diffraction analysis (XRD). It is a globular protein consisting in three domains folded around the cofactors. The 4Fe-4S cluster is coordinated by four cysteine residues and the iron atom of siroheme is coordinated by Cys 486. The cysteine rich domain, CRD, is high-conserved and essential for cofactor binding [29].

The reactions catalyzed by NR and NiR are schematized in Figure 2. The sequential reactions that permit to obtain ammonia from nitrate are globally a multi-electron reduction (totally eight electrons).

Geographical Region, Seasonal Differences and Fertilizers Use as Factors That Influence the Nitrate Content in Spinach

Extremely variable nitrate concentrations were found in spinach samples worldwide, varied from country to country, region to region, summer to winter crops and also conventional to organic cultivation. All these differences and variables imply a great deal of effort for scientific community to study, understand and construe the findings. EFSA Panel on Contaminant in Food Chain made an attempt of great relevance in 2008, providing a detailed trend of nitrate content in vegetables in Europe. Among almost 42.000 samples, approximately 7000 were spinaches coming from all Europe (North, South and Central). Only 5% of these samples resulted above the European legal limits (3500 and 2500 mg kg^{-1} for fresh and preserved/deep-frozen/frozen spinach, respectively) and the mean value of nitrate in Europe was 1066 mg kg^{-1}, significantly below the legal limits. Lower levels of nitrate were recorded in Mediterranean countries, while nitrate content may be higher in Central and North Europe. No significant seasonal difference was registered [30].

In order to verify, confirm and extend these data, we collected the most valuable and recent surveys and monitoring studies conducted on spinach samples all over the world. The results are graphically shown in Figure 3. More than 30 different researches were considered, conducted in 20 different countries. The mean value of nitrate content registered is 1653 mg kg^{-1}, with a global trend that seems to indicate a raise of nitrate accumulation in spinach. The European mean value (1544 mg kg^{-1}), obtained considering 9 countries (France, Spain, Portugal, Italy, Estonia, Belgium, Sweden, UK, Croatia), is slightly different from EFSA data of 2008, but in accordance with world rate [31-41, 14].

On the contrary, analysing USA rate, it can be noticed an inverse tendency from 2009 to 2015 [42, 43]. Similar studies coming from Asian continent also reported a wide range of nitrate levels in spinach with mixed and ambiguous trends [44-48]. Finally, lowest nitrate levels were registered in Iran and Turkey [49-53].

Although these evidences should be considered as indicator of nitrogen levels of spinach plants and soil, it must be pointed up the limitations of these surveys. In fact, apart comparing geographical provenience, many other parameters should be taken into consideration, such as production practices, harvest period and other operations that may lead to nitrate variations.

Analyzing in deep these surveys, similar trends can be verified. Several of these studies were focused on the seasonal differences. They proved that spinach crops grown in winter period have generally higher nitrate content compared to the summer crops. Gent reported that nitrate was 11% higher in winter than in other seasons [54, 55].

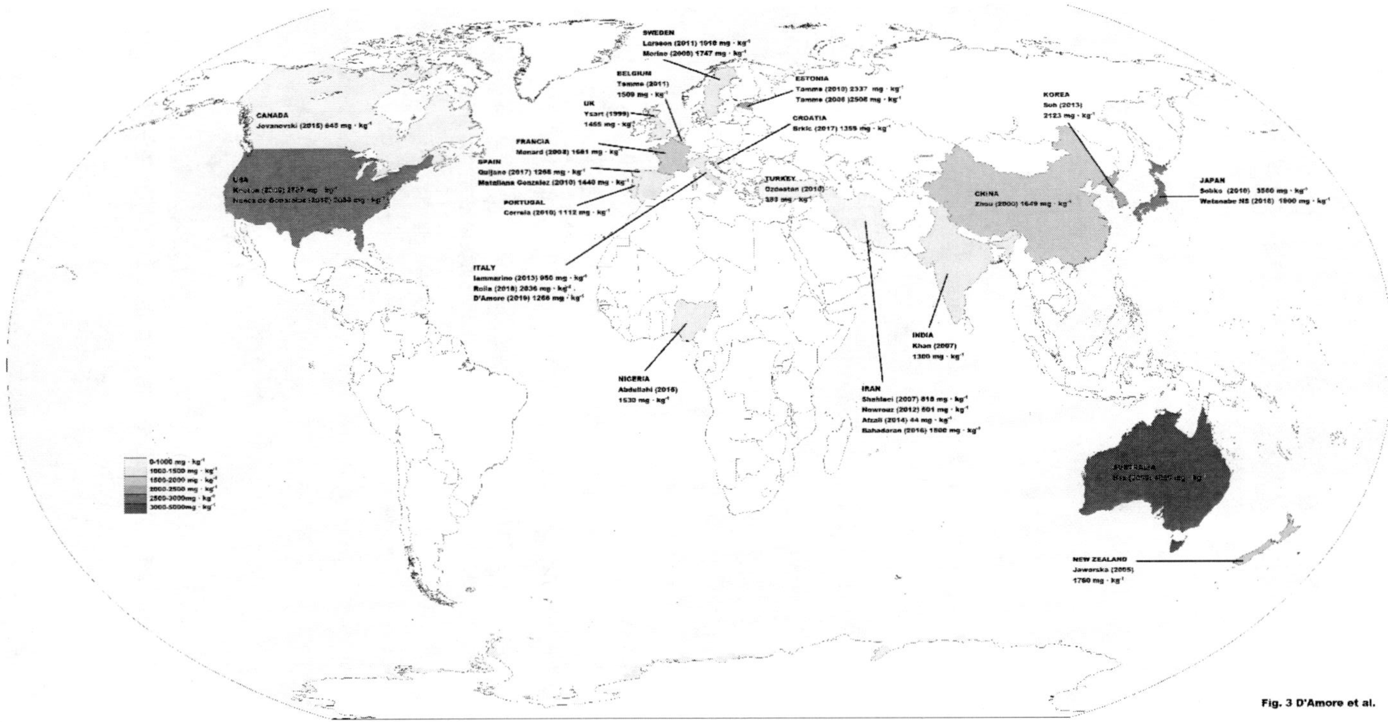

Figure 3. Results of monitoring studies of nitrate concentration in spinach worldwide. The nitrate amount is expressed in mg kg^{-1}.

Most of the mentioned data are in accordance, because favourable temperature and light intensity are important environmental factors that influence the nitrate metabolism of spinach. In fact, summer/spring spinach crops receive a light period higher than winter, so, as stated in the previous paragraph the pathway conversion $NO_3^-/NO_2^-/NH_4^+$/amino acids is more active, and the nitrate concentration in vacuoles is lower. The result is generally confirmed for spinach grown in greenhouse, indeed the nitrate concentration may be lower using an artificial lighting exposition period, due to maximum activity of NR. On the contrary, the variation in dark covered spinach is minimal. Some authors reported higher nitrate content in dark covered crops if compared to those grown in the open air during the same season.

Ozturk et al. advised that cloudy and foggy weather leads to the increase of nitrate concentration in spinach by interfering with the normal photosynthetic pathway. It is indicated that also desert climate or when water supply is cut off, can contribute to nitrate accumulation in spinach leaves [56].

On the contrary, a remarkable effect of climate on nitrate accumulation was observed by Anjana et al. 2007 in India; they indicated that during high rainfall period spinach harvest has a lower nitrate content, instead wet and dry climate leads to an increase of nitrate concentration [57].

Moreover, many interesting researches have been carried out regarding cultivation procedures. Studies have demonstrated lower nitrate content in spinach cultivated through organic agriculture if compared to conventionally grown spinach crops [58, 59]. Other studies reported the opposite [9], whereas other authors noted very few differences in NO_3^- concentration after analysis of samples labelled as "organic or conventional" in USA [41].

In order to provide a partial explanation of these overviews it should be underlined that organic fertilizers, that consist in manure, dried blood, green wastes, etc. slowly release nitrate in soil, with balanced absorption and no accumulation. These fertilizers are easily compostable and then used in organic agriculture. On the opposite, inorganic nitrogen fertilizers, also called mineral, (potassium nitrate, calcium ammonium nitrate, urea, ammonium nitrate and diammonium phosphate, etc.) directly provide nitrate to the soil, causing a fast increase of NO_3^- content in plant and an immediate activation of key genes transcription (*NRT1*, *NRT2*, *NIA1*, etc.). Also some herbicides and pesticides may produce nitrate as decomposition intermediate. In both cases over-fertilization practices, that favour nitrate accumulation in plants, should be avoided [2, 4, 8, 23].

Nitrite and Nitrate Toxicity

Nitrate and nitrate are potentially harmful if consumed in large amounts by humans. Although Nitrate is relatively non-toxic, due to its low reactivity, nitrite is more reactive [60]. The main raised concern is an acute syndrome occurring especially in infants < 6 months known as baby blue syndrome or infantile methemoglobinemia. Nitrite/nitrate ingested pass in blood circulation and react with the ferrous group of haemoglobin, which is oxidized to ferric state to form methemoglobin. Then, the oxygen cannot be transported to tissues, causing hypoxia and cyanosis. In infants and young children, the lack of the enzyme that catalyzes the reduction reaction of methemoglobin, represents an added risk factor [61].

Moreover, both, nitrate ingested and nitrate recirculated through salivary glands are in part converted to nitrite by commensal nitrate-reducing bacteria, located on the posterior part of the tongue [62, 63]. This conversion reaction may occur also in gastrointestinal tract due to the nitrate reductase enzyme, which is activated in the acid environment of stomach [64, 65]. The nitrite ingested or endogenous formed by nitrate reduction can react with secondary and tertiary amino-groups of proteins to form nitrosamines (Nitroso compounds - NOCs), a group of compounds classified by IARC (International Agency for Research on Cancer) as carcinogenic for humans. On the contrary, nitrate and nitrite were inserted in "2A group - probably carcinogenic to humans" and in 2015 IARC also declared the processed meat as "carcinogenic to humans – 1 group" [66, 67].

In particular, an increased risk of gastric, intestinal, oesophageal and pancreatic cancer was associated to high nitrate intake [68, 69]. An epidemiological observational study conducted in Korea correlated nitrate rich-diet with gastric cancer (GC) incidence, assigning a crucial role to carcinogenic products derived by cooking process. The researchers concluded "*Radish kimchi and cooked spinach were associated with increased risk of GC in our study*" [70].

Other researches, mainly carried out on mice, and several clinical studies demonstrated that nitrate may cause thyroid, teratogenic, metabolic toxic effects and brain cancer [71]. Many authors claim, is spite of these evidences, that these results are inconclusive [72].

On the other hand, it must be underlined that during last ten year many researchers have revaluated the nitrate impact on heath, showing some beneficial effects related to the conversion pathway of nitrate and nitrite to nitric oxide (NO) [73, 74]. NO is a signalling molecule, also called second messenger, which plays a key role in vasodilation, inhibition of blood platelets aggregation, regulation of immune response and inflammatory process. In fact, there is a class of drugs known as NO donors (nitroglycerin, nicorandil, isosorbide monitrate, etc.) that has therapeutic importance as antianginal, antihypertensive and heart failure medicinal products [75-77].

Since the NO_3^--NO_2^--NO pathway has been defined, several authors have proposed clinical applications in several pathological conditions. In particular, nitrate and nitrite supplements may produce a lot of benefits on cardiovascular system, such as lowering of blood pressure, maintenance of vessel homeostasis, improvement of endothelial function, protection against ischemia reperfusion injury and reduction in platelet aggregation [78, 79]. Furthermore, a protective role against gut pathogens due to antimicrobial effects of nitrate/nitrite salts was reported [80]. Finally, some authors have advised anti-obesity and anti-diabetic effects of nitrate and nitrite [81].

In recent times, the potential supplementation of nitrate-rich vegetables powders and juices was widely investigated. In nutraceutical field, spinach and beetroot powder and juice are becoming popular as dietary supplements or "functional foods." This is due to their capacity to improve vascular function, not only in pathological conditions, but also among healthy people and athletes during anaerobic exercise [82, 83, 84].

These data are in continuous evolution and the results lead to not univocal and uncertain conclusions. In fact, it appears necessary to develop new long-term studies and clinical trials, in order to clarify the potential risks and benefits deriving from a nitrate rich diet. Most likely, the intake of NO_3^- from vegetables-based diet may be considered sure due to the presence of protective compounds such as vitamin C (ascorbic acid), flavonoids, polyphenols and carotenoids which reduce NOCs formation enhancing nitrite reduction to NO [15, 85]. Indeed, different food sources and associations may result in different health adverse/beneficial effects. As far as said, this topic is worthy of researching and other risk assessment studies are needed.

Regulation and Principal Techniques for the Determination of Nitrite and Nitrate

Due to these potential remarkable negative effects, these contaminants are strictly monitored and their usage as food additives is regulated by different countries worldwide.

The first regulation that contains legal restriction of nitrate levels in water was approved in 1975 in by European Commission (Council Directive 75/440/EEC), as a results of some cases of methemoglobinaemia [86]. Later, in 1997 the maximum levels of nitrate in spinach and in other nitrate rich vegetables were set by European Member States (194/97) [87]. This regulation was amended in 2002 (EC No. 563/2002) and finally in 2011 (EC No. 1258/2011) [88, 89].

USA-FDA has never introduced maximum admissible levels for nitrate in vegetables. Instead, China Authorities have fixed this value to 3100 mg kg^{-1}.

Finally, regarding fresh meat and meat products the current regulation for NO_3^-/NO_2^- is the decision 1129/2011/EC [90].

In addition, the Joint Expert Committee of Food and Agriculture (JECFA – WHO/FAO Commission) and the Scientific Committee on Food (SCF - European Commission) established the acceptable daily intake (ADI) for contaminants and additives. These values were set at 3.7 and 0.06 mg kg b.w.$^{-1}$ for nitrate and nitrite, respectively, corresponding to 222 and 4.2 mg per day for a 60 kg adult (EU Scientific Committee for Food, 1995). These data were revaluated and confirmed both by WHO in 2002 and EFSA Panel on Contaminant in Food Chain in 2008 [30].

Instead, the US Environmental Protection Agency (EPA) set the maximum acceptable daily oral dose of toxic substances, known as Reference Dose (RfD), expressed in mg kg^{-1}day^{-1}. The assigned values for nitrate and nitrite are 1.6 and 0.1 mg kg^{-1} day^{-1}, respectively.

The analytical determination of nitrite and nitrate may be accomplished by using different techniques. The majority of the methods involve spectrophotometric procedures. In fact, a spectrophotometric determination, improved and modified during years, has become the reference method. Other determination methods include potentiometry, chemiluminescence, ion chromatography, polarography, capillary electrophoresis, HPLC and GC-MS. [84, 34, 91-95].

In recent years, the ion-exchange chromatography (IC) method with conductivity or UV detection has become the most adopted reference technique worldwide, due to its simplicity, rapidity and accuracy [96, 97].

Materials and Methods

Experimental

Sampling and Chemicals

30 spinaches samples (14 fresh, 13 fresh-IV range and 13 frozen) were collected from local stores during the period November 2018 to February 2019. The samples were stored under refrigeration (4°C ± 2°C) until analysis. The samples were originated from 7 Italians regions (Apulia: 33%; Lazio: 7%; Emilia-Romagna: 20%; Campania: 10%; Abruzzo: 10%;

Piedmont: 3,5%; Basilicata: 16.5%). Two replicates of each sample were analyzed and nitrates and nitrites contents were evaluated as the mean of two measurements.

Nitrites and nitrates certified anion standard solutions for ion chromatography (1000 mg L^{-1}) were supplied from Sigma-Aldrich (Stenheim, Germany). Carbonate-free sodium hydroxide (50%, w/w) and anhydrous sodium carbonate (>99.5%) were purchased by J.T. Baker (Deventer, Netherlands). Ultrapure water with a specific resistance of 18.2 MΩ-cm, supplied by Milli-Q RG unit from Millipore (Bedford, MA, USA), was used for sample extraction and for mobile phases preparation. The mobile phases were filtered with a 0.45 μm membrane and degassed with helium. From nitrites and nitrates standard solutions at concentration of 1000 mg L^{-1}, the following diluted concentrations were used: 0.2, 0.4, 0.8, 1.6, 6.5 mg L^{-1} for nitrites and 0.6, 1.2, 2.4, 4.8, 12.5 mg L^{-1} for nitrates. Ultrapure water was used for dilutions.

Sample Preparation

A 100-g portion of the sample edible part was homogenized by using a mixer (BUCHI B-400, BUCHI Italia s.r.l., Assago, Milan, Italy). 4g of homogenized sample was mixed with 200 mL of ultrapure water and then placed at 70°C for five minutes. After cooling, the mixture was filtered using Whatman No. 41, 150 mm filters (Whatman, Springfield Mill, UK) and then ~ 2-3 mL of the filtrate were filtered through Anotop 10 LC, 0.2 μm, 10 mm filters (Whatman, Springfield Mill, UK) prior to chromatographic analysis. For nitrate concentrations higher than 625 mg kg^{-1} a dilution with ultrapure water was necessary to fall within the measurement range.

Apparatus and Method

All the chromatographic analysis were performed by using a Dionex system (Dionex Corporation, Sunnyvale, CA) composed of a GP50 quaternary gradient pump, a Rheodyne injection valve (model RH9125, Cotati, CA, USA) with a 25 μL injection loop. A Dionex anion self-regenerating suppressor (ASRS II, 4 mm) at operative current of 50 mA and an electrochemical detector set to conductivity mode equipped with a temperature compensated conductivity cell (model ED40) were used. The plastic reservoir bottles (DX500 2L bottles, Dionex) were pressurized with pure Nitrogen to 0.8 MPa. The system was interfaced via proprietary network chromatographic software (PeakNetTM, Dionex Corporation, Sunnyvale, CA) to a personal computer for instrumentation control, data acquisition and processing. The IonPac® AS9-HC column (250mm×4mm i.d., particle size: 13μm) coupled with an AG9-HC precolumn (Dionex Corporation, Sunnyvale, CA) was used for chromatographic separations. A 9 mM sodium carbonate solution was used as mobile phase in isocratic mode and a flow rate of 1.0 mL min^{-1}. Total run time: 20 minutes.

The analytical method used for determining nitrites and nitrates in spinaches was developed from the Official Method EN 12014-4:2005, optimizing a chromatographic separation proposed by Dionex Corporation for the determination of inorganic anions [96, 98].

Table 2. Validation parameters of two analytical methods

Parameter	**HPIC***		**CIC****	
	NO_3^-	**NO_2^-**	**NO_3^-**	**NO_2^-**
Linearity (r^2)	0,999	0,999	0.999	0.999
LOD***	3.2	1.5	1.2	0.7
LOQ****	9.6	4.5	3.7	2.1
Specificity	Confirmed for meat and dairy products, leafy vegetables, seafood and animal feeds			
Mean precision (%) (n=18)	4.2	4.1	2.7	3.4
Mean recovery (%) (n=18)	98.3	98.7	100.8	98.8
Measurement uncertainty (%)	9.7	7.8	2.4	2.6
Robustness Application field	Meat and dairy products, leafy vegetables, seafood and animal feeds			

*HPIC: High pressure Ion Chromatography.
**CIC: Capillary Ion Chromatography.
***LOD: limit of detection (expressed in mg kg^{-1} of sample).
****LOQ: limit of quantification (expressed in mg kg^{-1} of sample).

This method was validated following an *in-house* validation model according to Thompson Harmonized Validation Guidelines, in agreement with Decision 657/2002/EC [99, 100]. The most important validation parameters of this method are reported in Table 2. This analytical method was accredited by the Italian Organism for laboratory accreditation ACCREDIA and it is used routinely within official controls at Chemistry Department of Experimental Zooprophylactic Institute of Apulia and Basilicata (Foggia, Italy).

In order to confirm the most interesting data obtained during this monitoring, in terms of both nitrate and nitrite, the samples characterized by nitrate concentration higher than the legal limit and/or by nitrite concentration well higher than the method limit of quantification were analyzed by using an innovative analytical method by capillary ion chromatography [101]. Both ion chromatography techniques adopted in this survey were successfully involved in international studies aimed at certifying reference materials of spinach powder for nitrate determination [93].

These analysis were carried out by using a Dionex ICS-4000 capillary HPIC system (Thermo Fisher Scientific, Waltham, USA), composed of a 4-port high-pressure valve with a 0.4 μL internal sample loop, a dual-stepper motor pump, an EGC KOH eluent generator cartridge, an EG degas cartridge, a column heater set at 15°C, an ACES™ 300 anion capillary electrolytic suppressor cartridge, set at 10 mA, and a conductivity detector. The chromatographic separations were accomplished by using a Dionex IonPac® AS11-HC capillary column (250 mm x 0.4 mm i.d.; Particle size, 9 μm: Thermo Fischer Scientific, Waltham, USA). The elution gradient was based on KOH: isocratic step with 2mM KOH for 9 min, gradient step to 15 mM in 1 min, another gradient step up to 30 mM in 26 min, then the system was brought back to 2 mM in 1 min and then re-equilibrated at this concentration for 1 min (total run time: 38 min; flow rate: 0.015 mL min^{-1}). The system was interfaced via proprietary network chromatographic software (Chromeleon 7, Thermo Fisher Scientific, Waltham, USA) to a personal computer for instrumentation control, data acquisition and processing. Also, this method was fully validated and the most important validation parameters of this method are reported in Table 2.

RESULTS AND DISCUSSION

The results obtained during this monitoring are reported in Table 3. As expected, nitrates are present at high levels in spinaches, up to 3553.5 ± 344.7 mg kg^{-1} registered in a fresh spinaches sample (IV range), and the mean value measured in 30 samples was equal to 1266.0 mg kg^{-1}. In three spinaches samples (10.0% of total samples analyzed) (1 IV range and 2 frozen), concentrations higher than the legal limit (3500 and 2000 mg kg^{-1} for IV range and frozen spinaches, respectively) were verified. These concentrations resulted equal to 3553.5 ± 344.7, 2181.3 ± 211.6 and 2337.1 ± 226.7 mg kg^{-1}.

Quantifiable amounts of nitrites (higher than method LOQ = 4.5 mg kg^{-1}) were detected in 10 samples (33.3% of total samples analyzed). The presence of nitrite in leafy vegetables is not admitted by the actual Normative, so, this result is significant. Nitrites residues were detected in fresh (IV range) and in frozen spinaches, 6 and 4 samples, respectively, at levels varying from 10.0 ± 0.8 to 78.5 ± 6.1 mg kg^{-1}. No nitrite residues were detected in fresh spinaches samples.

Table 3. Monitoring results of 30 spinach samples

Type of sample	Number of analyzed samples	Number of samples with nitrites concentrations higher than method LOQ	Nitrites concentration range and mean concentration*	Nitrates concentration range and mean concentration*	Number of samples with nitrates concentration higher than legal limits
Fresh spinaches	14	0	-	75.5 – 1846.9 (1137.6)	0
Fresh spinaches (IV range)	13	6	10.0 – 78.5 (47.5)	577.3 – 3553.5 (1559.0)	1 (3553.5)*
Frozen spinaches	13	4	10.0 – 47.0 (34.2)	242.7 – 2337.1 (1111.3)	2 (2181.3 – 2337.1)*

* Expressed as mg kg^{-1} in the sample.

Although the mean nitrate concentration obtained in fresh IV range spinaches (1559.0 mg kg^{-1}) resulted higher that those registered in fresh and frozen spinach samples (1137.6 and 1111.3 mg kg^{-1}, respectively), this difference resulted as not significant ($p \leq 0.05$). Indeed, by analyzing the differences among the concentrations obtained for three type of spinaches collected (fresh, IV range and frozen) through ANOVA one-way test ($p \leq 0.05$), it is not possible to verify a statistically significant difference.

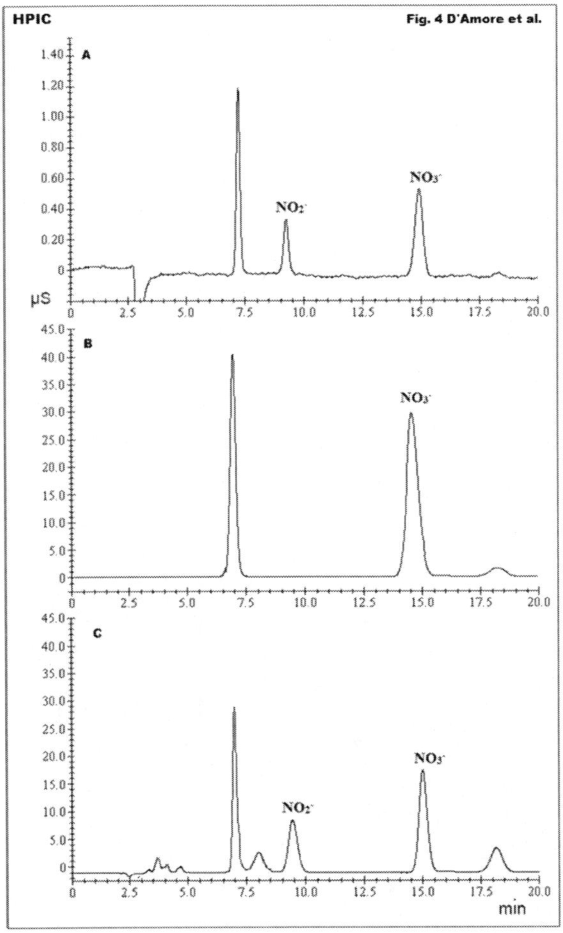

Figure 4. Chromatograms of two "not compliant" spinach samples, obtained by HPIC: (A) Standard solution of nitrites and nitrates (2.0 mg L^{-1}); (B) Frozen spinach sample, $C_{Nitrate}$ = 2291.8 mg kg^{-1}; (C) Fresh spinach sample - IV range, $C_{Nitrite}$ = 78.5 mg kg^{-1}.

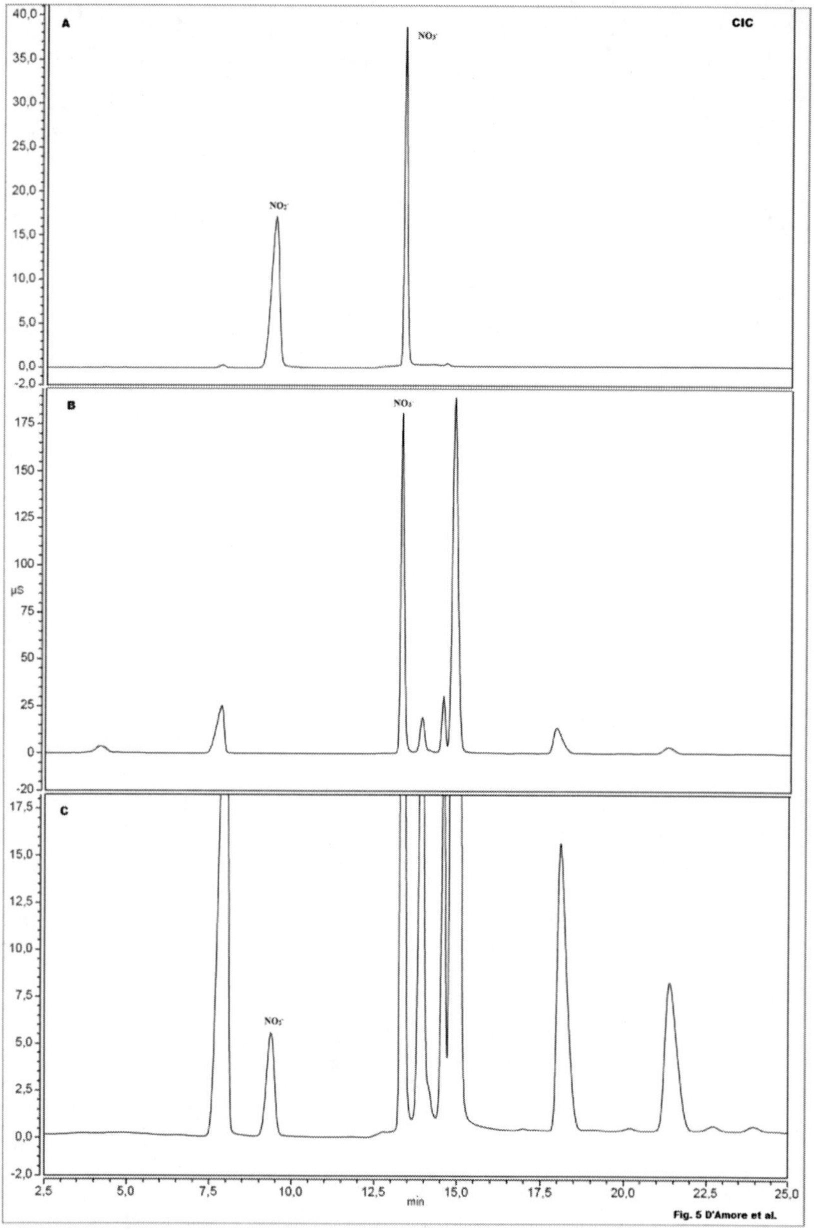

Figure 5. Chromatograms of two "not compliant" spinach samples, obtained by CIC: (A) Standard solution of nitrites and nitrates (20.0 mg L^{-1}); (B) Frozen spinach sample, $C_{Nitrate}$ = 2291.8 mg kg^{-1}; (C) Fresh spinach sample – IV range, $C_{Nitrite}$ = 78.5 mg kg^{-1}.

In order to confirm the most significant results of this survey, from a point of view of analytical determinations, both the samples with nitrates levels higher than the legal limit (3 samples) and those with quantifiable nitrite concentrations (10 samples) were analyzed by using the "confirmatory" technique. The results obtained by ion chromatography were confirmed, since the concentrations detected by using the second technique resulted as comparable, that is included in the method measurement uncertainty (7.8% and 9.7% for nitrite and nitrate, respectively).

In Figures 4 and 5, two chromatograms of a "not-compliant" spinaches sample containing 2291.8 mg kg^{-1} of nitrates (4B and 5B) and of a spinaches sample containing 78.5 mg kg^{-1} of nitrites (4C and 5C) are shown. The chromatograms obtained by using both techniques are shown.

Very interesting comments related to both nitrite and nitrate can be made if this study is compared to a similar one carried out in 2014, in which 75 spinaches samples (21 fresh, 18 IV range and 36 frozen) were analyzed by using the same ion chromatography method [34]. Indeed, the most significant results, related to nitrates higher than the legal limit and to nitrite quantifiable levels are confirmed.

Fresh spinaches may be considered as "safest" among different types of products available on the market. Indeed, nitrite quantifiable levels were detected only in IV range and frozen spinaches. It is possible to suppose that a particular step in spinaches processing (i.e., washing, cut, etc.), used for obtaining both the IV range and the frozen products, may cause the nitrate reduction to nitrite. Future studies will be focused on deepening this aspect. Regarding high nitrate levels, no concentrations higher than the legal limit were detected in fresh spinaches samples. Regarding frozen spinaches, the probability to obtain a "not complaint" sample is higher than fresh and IV range, due to the lower legal limit (2000 for frozen, 3500 for fresh and IV range). This is due to the release of nitrates from the product after thawing out that should decrease the residual level in the product, substantially. However, this decrease could be not such high, and then the residual levels may exceed 2000 mg kg^{-1}. The nitrate concentrations higher than the legal limit registered in IV range samples could be due to the nitrate accumulation in the leaf that can proceed after packaging, during the entire shelf-life of the product. Also in this case future studies will be developed to fill this gap in knowledge about this matter.

As a further consideration, taking into account the not negligible percentage of "not-compliant" samples obtained, relating to nitrates concentration (10.0%), control is needed. Moreover, it is possible to affirm that nitrite residues may be present in IV range and frozen spinaches. Therefore, it is suggested to evaluate and establish a "maximum admissible level" for nitrites related to these types of spinaches for integrating the Communities Regulations.

CONCLUSION

In this chapter, we have described the accumulating pathway of nitrate/nitrite in spinach plants and the phytochemical and toxicological aspects of these two ions.

Spinach belongs to nitrate-rich vegetables. For this reason, maximum permitted values were set and many monitoring studies were developed worldwide.

A specific monitoring was carried out in this work for evaluating the levels of nitrate/nitrite in different types of spinach (30 samples of fresh, fresh (IV range) and frozen spinach), by using a validated ion chromatography method. The most interesting results were confirmed by using a validated capillary ion chromatography method. The conductivity detection was used in both techniques.

The results confirmed the findings reported about this topic until now. Fresh spinaches may be considered as "safest" among different types of products available on the market. This is due to both nitrite ion that was quantified only in IV range and frozen spinaches and nitrate levels that resulted higher than legal limits only in one sample of fresh-IV range and two samples of frozen spinaches.

The reason of these high levels of nitrite/nitrate in IV range and frozen spinaches has to be deepened in future studies. It is possible to suppose that a particular step in spinaches processing (i.e., washing, cut, etc.), used for obtaining both the IV range and the frozen products, may cause the nitrate reduction to nitrite. Regarding nitrate, the concentrations higher than the legal limit registered in IV range samples could be due to the nitrate accumulation in the leaf that can proceed after packaging.

A final comment may be stated regarding the not negligible percentage of "not-compliant" samples for nitrate (10.0%). Indeed, it is possible to confirm that control is needed. Moreover, the possible presence of quantifiable residues of nitrite in IV range and frozen spinaches was confirmed. This last result allows suggesting the introduction of a specific "maximum admissible level" for nitrites in these types of spinaches, for integrating the Communities Regulations.

REFERENCES

[1] Bernhard, A. (2010) The Nitrogen Cycle: Processes, Players, and Human Impact. *Nature Education Knowledge,* 3(10):25.

[2] Nasholm, T., Kielland, K. and Ganeteg, U. (2009). Uptake of organic nitrogen by plants. *New Phytologist,* 182: 31-48.

[3] Stevens, C. J. (2019). Nitrogen in the environment. *Science,* 363: 578-580.

[4] Iammarino, M. and Di Taranto, A. (2014) Nitrogen fertilizers and nitrite-nitrate accumulation in foodstuffs: A survey. In *Nitrogen Fertilizer: Agricultural Uses, Management Practices and Environmental Effects*, Nova Publisher, Karlene Winslow, pp.1-29.

[5] EFSA (2017) Panel on food additives and nutrient sources added to food, Re-evaluation of potassium nitrite (E 249) and sodium nitrite (E 250) as food additives. *The EFSA Journal*, 15(6).

[6] Iammarino, M., Di Taranto, A., Cristino, M. (2013) Endogenous levels of nitrites and nitrates in wide consumption foodstuffs: Results of five years of official controls and monitoring. *Food Chemistry,* 140(4) 763-71.

[7] Keeton, J. T. (2011) History of nitrite and nitrate in food. In *Nitrite and Nitrate in Human Health and Disease*, edited by Nathan Bryan and Joseph Loscalzo, 69–84.

[8] Gaskell, M. and Smith, R. (2007): Nitrogen sources for organic vegetable crops. *Hort Technology*, 17:431-441.

[9] Ranasinghe R. A. S. N. and Marapana R. A. U. J. (2018) Nitrate and nitrite content of vegetables: A review. *Journal of Pharmacognosy and Phytochemistry,* 7(4): 322-328.

[10] Colla, G., Kimb, H., Kyriacouc, M. C., Rouphaeld, Y. (2018) Nitrate in fruits and vegetables. *Scientia Horticulturae,* 237 221–238.

[11] Santamaria, P. (2006). Nitrate in vegetables: Toxicity, content, intake and EC regulation. *Journal of the Science of Food and Agricolture,* 86:10–17.

[12] Cintya, H., Silalahi, J., De Lux, P., Siburian, R. (2018) The influence of storage condition on nitrite, nitrate and vitamin C levels in vegetables. *F1000Research,* 7.

[13] Chung, J. C, Chou, S. S., Hwang, D. F. (2004) Changes in nitrate and nitrite content of four vegetables during storage at refrigerated and ambient temperatures. *Food Additives and Contaminants,* 21(4):317-322.

[14] Brkić, D., Bošnir, J., Bevardi, M. et al., (2017) Nitrate in leafy green vegetables and estimated intake. *African Journal of Traditional, Complementary, and Alternative Medicines*, vol. 14, no. 3, pp. 31–41.

[15] Santamaria, P., Elia, A., Serio, F. and Todaro, E. (1999) A survey of nitrate and oxalate content in retail fresh vegetables. *Journal of the Science of Food and Agricolture,* 79:1882–1888.

[16] Miller, A. J. and Cramer, M.D. (2005) Root nitrogen acquisition and assimilation. *Plant and Soil* 274, 1–36.

[17] Tischner, R. and Kaiser, W. (2007) Nitrate Assimilation in plants. *Biology of nitrogen Cycle.*

[18] Forde, B. G. and Walch-liu, P. (2009) Nitrate and glutamate as environmental cues for behavioural responses in plant roots. *Plant, Cell and Environment,* 32, 682–693.

[19] Bose, B., Srivastava, H.S., (2001) Absorption and accumulation of nitrate in plants: influence of environmental factors. *Indian Journal of Experimental Biology,* 39, 101–110.

[20] Wang, X., Cai, X., Xu, C., Wang, S., Dai, S., Wang, Q. (2018). Nitrate Accumulation and Expression Patterns of Genes Involved in Nitrate Transport and Assimilation in Spinach. *Molecules,* 23(9), 2231.

[21] Harrison, J., Hirel, B. and Limani, A. M. (2004). Variation in nitrate uptake and assimilation between two ecotypes of *Lotus japonicus* and their recombinant inbred lines. *Physiologia Plantarum*, 120: 124-131.

[22] Ott, K., Koenig, R., Miles, C. (2008) Influence of Plant part on Nitrate Concentration in Lettuce and Spinach. *International Journal of Vegetable Science,* 14-4:351-361.

[23] Chen, B. M., Wang, Z. H., Li, S. X., Wang, G. X., Song, H. X., Wang, X. N. (2004) Effects of nitrate supply on plant growth, nitrate accumulation, metabolic nitrate concentration and nitrate reductase activity in three leafy vegetables. *Plant Science*, 167: 635–643.

[24] Kaiser, W.M. and Brendle-Behnisch E. (1991) Rapid modulation of spinach leaf nitrate reductase activity by photosynthesis I. Modulation in vivo by CO_2 availability. *Plant Physiology,* 96, 363-367.

[25] Samuolienė, G., Urbonavičiūtė, A., Duchovskis, P., Bliznikas, Z., Vitta, P. and Žukauskas, A. (2009). Decrease in nitrate concentration in leafy vegetables under a solid-state illuminator. *Horticultural Science,* 44, 1857–1860.

[26] Riens, B., Heldt, H.W., (1992) Decrease of nitrate reductase activity in spinach leaves during a light-dark transition. *Plant Physiology,* 98, 573–577.

[27] MacKintosh, C. (1992) Regulation of spinach-leaf nitrate reductase by reversible phosphorylation. *Biochimica et Biophysica Acta.* 6; 1137(1):121-6.

[28] Weiner, H., Kaiser, W. M. (1999) 14-3-3 proteins control proteolysis of nitrate reductase in spinach leaves. *FEBS Letters,* 455 75–78.

[29] Swamy, U., Wang, M., Tripathy, J.N., Kim, J., Hirasawa, M., Knaff, D. B., Allen, P. (2005) Structure of Spinach Nitrite Reductase: Implications for Multi-electron Reactions by the Iron–Sulfur: Siroheme Cofactor. *Biochemistry,* 44 (49).

[30] EFSA (2008) Nitrate in vegetables: Scientific opinion of the panel on contaminants in food chain. *The EFSA Journal,* 689: 1-79.

[31] Correia, M., Barroso, Â., Barroso, M.F., Soares, D., Oliveira, M. B. P. P., Delerue-Matos, C. (2010) Contribution of different vegetable types to exogenous nitrate and nitrite exposure. *Food Chemistry,* 120, 960-966.

[32] Matallana González, M. C., Martínez-Tomé, M. J. and Torija Isasa, M. E. (2010) Nitrate and nitrite content in organically cultivated vegetables. *Food Additives and Contaminants*: Part B, 3:1, 19-29.

[33] Menard, C., Heraud, F., Volatier, J. L. and Leblanc J. C. (2008) Assessment of dietary exposure of nitrate and nitrite in France. *Food Additives and Contaminants*, 25:8, 971-988.

[34] Iammarino, M. and Di Taranto, A. (2014) Monitoring of nitrites and nitrates levels in leafy vegetables (spinach and lettuce): a contribution to risk assessment. *Journal of the Science of Food and Agricolture,* 94,4.

[35] Temme, E. H., van devijvere, S., et al. (2011) Average daily nitrate and nitrite intake in the Belgian population older than 15 years. *Food Additives and Contaminants Part A,* 28:1193–204.

[36] Larsson, K., Darnerud, P. O., Ilbäck, N. G., Merino, L. (2011) Estimated dietary intake of nitrite and nitrate in Swedish children. *Food Additives and Contaminants,* 28: 659-666.

[37] Quijano, L., Yusà, V., Font, G., McAllister, C., Torres, C., Pardo, O. (2016) Risk assessment and monitoring programme of nitrates through vegetables in the Region of Valencia (Spain). *Food Chemistry and Toxicology,* 100:42-49.

[38] Tamme, T., Reinik, M., Roasto, M., Meremae, K., Kiis, A. (2010). Nitrate in leafy vegetables, culinary herbs, and cucumber grown under cover in Estonia: content and intake. *Food Additives and Contaminants: Part B,* 3(2):108-113.

[39] Tamme, T., Reinik, M., Roasto, M., Juhkam, K., Tenno, T., Kiis, A. (2006) Nitrates and nitrites in vegetables and vegetable-based products and their intakes by the Estonian population. *Food Additives and Contaminants,* 23: 355–361.

[40] Roila, R., Branciari, R., Staccini, B., Ranucci, D., Miraglia, D. M., Haouet, N. (2018) Contribution of vegetables and cured meat to dietary nitrate and nitrite intake in Italian population: Safe level for cured meat and controversial role of vegetables. *Italian Journal of Food Safety,* 7:76-92.

[41] Ysart, G., Miller, P., Barrett, G., Farrington, D., Lawrance, P., Harrison, N. (1999) Dietary exposures to nitrate in the UK. *Food Additives and Contaminants*, 16, 521-532.

[42] Keeton, J., Osburn, W., Hardin, M., Longnecker, M., Bryan, N. (2009). *A national survey of the nitrite/nitrate concentrations in cured meat products and non-meat foods available at retail.* American Meat Institute Foundations.

[43] Nuñez de González, M. T., Osburn, W. N., Hardin, M. D., Longnecker, M., Garg, H. K., Bryan, N. S., Keeton, J.T. (2015). A survey of nitrate and nitrite concentrations in conventional and organic-labeled raw vegetables at retail. *Journal of Food Science,* 80(5):C942-9.

[44] Suh, J., Jin Paek ,O., Kang, Y., Ahn, J.E., Jung, J. S., An, Y. S., Park, S. H. (2013) Risk Assessment on Nitrate and Nitrite in Vegetables Available in Korean Diet. *Journal of Applied Biological Chemistry,* 56, 205-211.

[45] Zhou, Z.,Wang, M.,Wang, J. (2000) Nitrate and nitrite contamination in vegetables in China. *Food Reviews International,* 16:61–76.

[46] Khan, M. A. I., Ueno, K., Horimoto, S., Komai, F., Tanaka, K. and Ono Y., (2007) Evaluation of the physio-chemical and microbial

properties of green tea waste– rice bran compost and the effect of the compost on spinach production. *Plant Production Science,* 10.

[47] Sobko, T., Marcus, C., Govoni, M. and Kamiya, S. (2010) Dietary nitrate in Japanese traditional foods lowers diastolic blood pressure in healthy volunteers. *Nitric Oxide,* 22(2):136 40.

[48] Watanabe, N.S., Yamasaki, H. (2016) Dynamics of Nitrite Content in Fresh Spinach Leaves: Evidence for Nitrite Formation Caused by Microbial Nitrate Reductase Activity. *Journal of Nutrition and Food Sciences,* 7: 572.

[49] Bahadoran, Z. and Mirmiran, P. (2016) Nitrate and nitrite content of vegetables, fruits, grains, legumes, dairy products, meats and processed meats. *Journal of Food Composition and Analysis,* 51, 93-105

[50] Ozdestan, Z. and Uren, A. (2010) Development of a Cost-Effective Method for Nitrate and Nitrite Determination in Leafy Plants and Nitrate and Nitrite Contents of Some Green Leafy Vegetables Grown in the Aegean Region of Turkey. *Journal of Agricultural and Food Chemistry,* 58, 5235–5240.

[51] Nowrouz, P., Taghipour, H., Dastgiri, S., Bafandeh, Y., Hashemimajd, K. (2012) Nitrate Determination of Vegetables in Varzeghan City, Northwestern Iran. *Health Promotion Perspectives,* (2): 244-250.

[52] Shahlaei, A., Ansari, N. A., Dehkordie, F. S. (2007) Evaluation of Nitrate and Nitrite Content of Iran Southern (Ahwaz) Vegetables during Winter and Spring of 2006. *Asian Journal of Plant Sciences,* 6: 1197-1203.

[53] Afali, S. F., Elahi, F. (2014) Measuring nitrate and nitrite concentrations in vegetables, fruits in Shiraz. *Journal of Applied Science and Environmental Management,* 18 (3) 451-4.

[54] Gent, M. P. N. (2016) Effect of irradiance and temperature on composition of spinach. *Horticultural Science,* 51:133–140.

[55] Jaworska, G. (2005) Content of nitrates, nitrites and oxalates in New Zealand spinach. *Food Chemistry,* 89, 235–242.

[56] Ozturk, M. G., Salih, S., Mehmet, B., Süleyman, B. (2013) Nitrate and Edible Plants in the Mediterranean Region of Turkey: An Overview. In *Nitrate in Leafy Vegetables*, International Publishing House Umar 17-51.

[57] Anjana, S. U. and Iqbal, M. (2007) Nitrate accumulation in plants, factors affecting the process, and human health implications. A review. *Agronomy for Sustainable Development*, 27: 45-57.

[58] Abubaker, S., Abu-Zahra, T., Alzu'bi, Y., Ammari, T., Tahboub, A. (2010) Nitrate accumulation in spinach (Spinacia oleracea L.) tissues under different fertilization regimes. *Journal of Food, Agriculture and Environment,* 8; 778-780.

[59] Gorenjak, A.H. (2013) Nitrate in vegetables and their impact on human health. A review. *Acta Alimentaria,* 42 (2):158-172.

[60] Parvizishad, M., Dalvand, A., Mahvi, A. H., Goodarzi, F., (2017) A Review of Adverse Effects and Benefits of Nitrate and Nitrite in Drinking Water and Food on Human Health. *Health Scope,* 14 164.

[61] Phillips, W. E. J. (1971) Naturally occurring nitrate and nitrite in foods in relation to infant methaemoglobinaemia. *Food and Cosmetics Toxicology,* 9:219-228.

[62] Qu, X. M., Wu, Z.F., Pang, B.X., Jin, L.Y., Qin, L. Z., Wang, S. L. (2016) From nitrate to nitric oxide: The role of salivary glands and oral bacteria. *Journal of Dental Researches,* 95, 1452–1456.

[63] Ferreira, R.M., Pereira-Marques, J., Pinto-Ribeiro, I., Costa, J. L., Carneiro, F., Machado, J. C., Figueiredo, C. (2018) Gastric microbial community profiling reveals a dysbiotic cancer-associated microbiota. *Gut,* 67 226-236.

[64] Hyeong, S. K., Sun, J. H. (2017) Changes of sodium nitrate, nitrite, and N-nitrosodiethylamine during in vitro human digestion. *Food Chemistry*, 225, 197-201.

[65] Lundberg, J. O., Carlstrom, M., Weitzberg, E. (2018) Metabolic Effects of Dietary Nitrate in Health and Disease. *Cell Metabolism,* 28 9-22.

[66] International Agency for Research on Cancer - IARC Working Group (2015) Consumption of red meat and processed meat. *IARC Monographs of Evaluation of Carcinogenic Risks to Humans,* 114.

[67] Gangolli, S. D., van den Brandt, P. A., Feron, V. J., Janzowsky, C., Koeman, J. H., Speijers, G. J., Spiegelhalder, B., Walker, R., Wisnok, J. S. (1994) Nitrate, nitrite and N-nitroso compounds. *Environmental Toxicology and Pharmacology,* 292(1) 1-38.

[68] Bedale, W., Sindelar, J. J., Milkowski, A. L. (2016) Dietary nitrate and nitrite: Benefits, risks, and evolving perceptions. *Meat Science,* 120 85-92.

[69] Ziarati, P. (2018) Potential Health Risks and Concerns of High Levels of Nitrite and Nitrate in Food Sources. *SciFed Pharmaceutics Journal,* 1:3.

[70] Kim, H. J., Chang, W. K., Kim, M. K., Lee, S. S., and Choi, B. Y. (2002) Dietary factors and gastric cancer in Korea: A case-control study. *International Journal of Cancer,* 97, 531–535.

[71] Hord, G. N., Tang, Y., Bryan, N. S. (2009) Food sources of nitrates and nitrites: the physiologic context for potential benefits. *The American Journal of Clinical Nutrition,* 1-10.

[72] Norman, G., Conley, M. N. (2017) Regulation of Dietary Nitrate and Nitrite: Balancing Essential Physiological Roles with Potential Health Risks. In *Nitrite and Nitrate in Human Health and Disease*, N.S. Bryan, J. Loscalzo (Eds.), Humana Press, Cham, pp. 153-162.

[73] Bondonno, C. P., Liu, A. H., Croft, D., Ward, N. C., Puddey, I. B., Woodman, R. J., Hodgson, J. M. (2015) Short-Term Effects of a High Nitrate Diet on Nitrate Metabolism in Healthy Individuals. *Nutrients*, 7, 1906-1915.

[74] Chamandoost, S., Moradi, M. F., Hosseini, M. J. (2016) A Review of Nitrate and Nitrite Toxicity in Foods. *Journal of Human Environment and Health Promotion,* 1 (2):80-86.

[75] Ferrari, R., Camici, P. G., Crea, F., Danchin, N., Fox, K., Maggioni, A. P., Manolis, A. J., Marzilli, M., Rosano, G. M. C., Lopez-Sendon, J. L. (2018) A 'diamond' approach to personalized treatment of angina. *Nature Reviews Cardiology,* 15, 120–132.

[76] Bath, P. M. W, Krishnan, K., Appleton, J. P. (2017) Nitric oxide donors (nitrates), L-arginine, or nitric oxide synthase inhibitors for acute stroke. *Cochrane Database of Systematic Reviews*, Issue 4.

[77] Jackson, J., Patterson, A. J., MacDonald-Wicks, L. and McEvoy M. (2017) The role of inorganic nitrate and nitrite in CVD. *Nutrition Research Reviews*, 30, 247–264.

[78] Jovanovski, E., Bosco, L., Khan, K., Au-Yeung, F., Ho, H., Zurbau, A., Jenkins, A.L., Vuksan, V. (2015) Effect of Spinach, a High Dietary Nitrate Source, on Arterial Stiffness and Related Hemodynamic Measures: A Randomized, Controlled Trial in Healthy Adults. *Clinical Nutrition Research,* 4(3):160-167.

[79] Jonvik, K. L., Nyakayiru, J., Pinckaers, P. J, M,, Senden, J. M. G., van Loon, L. J. C., Verdijk, L. B. (2016) Nitrate-rich vegetables increase plasma nitrate and nitrite concentrations and lower blood pressure in healthy adults. *Journal of Nutrition,* 146:986–93.

[80] Lundberg, J. O., Weitzberg, E., Cole, J. A., Benjamin, N. (2004) Nitrate, bacteria and human health. *Nature Reviews Microbiology,* 2:593–602.

[81] Ghasemi, A. and Jeddi, S. (2017) Anti-obesity and anti-diabetic effects of nitrate and nitrite. *Nitric Oxide*, 70 (1), 9–24.

[82] Salehi, B., Tumer, T., Özleyen, A., Peron, G., Dall'Acqua, S. et al. (2019) Plants of the Genus Spinacia: from bioactive molecules to food and phytopharmacological applications. *Trends in Food Science & Technology,* 88:260-273.

[83] Lidder, S., Webb, A. J. (2013) Vascular effects of dietary nitrate (as found in green leafy vegetables and beetroot) via the nitrate-nitrite-nitric oxide pathway. *British Journal of Clinical Pharmacology,* 75, 677–696.

[84] Shah, I., Petroczi, A., James, R. A., Naughton, D.P. (2013) Determination of Nitrate and Nitrite Content of Dietary Supplements Using Ion Chromatography. *Journal of Analytical Bioanalytical Techniques,* S12: 003.

[85] Menezes, F. E., Peixoto, L. G., Teixeira, R. R., Justino, A., Puga, G. M., Espindola, F. (2019) Potential Benefits of Nitrate

Supplementation on Antioxidant Defense System and Blood Pressure Responses after Exercise Performance. *Oxidative Medicine and Cellular Longevity*, 1-10.

[86] European Union (1975) Council Directive 75/440/EEC of 16 June 1975 concerning the quality required of surface water intended for the abstraction of drinking water in the Member States.

[87] European Union (1997) Commission Regulation (EC) No 194/97 of 31 January 1997 setting maximum levels for certain contaminants in foodstuffs.

[88] European Union (2002) Commission Regulation (EC) No 563/2002 of 2 April 2002 amending Regulation (EC) No 466/2001 setting maximum levels for certain contaminants in foodstuffs.

[89] European Union (2011) Commission Regulation (EU) No. 1258/2011 amending Regulation (EC) No. 1881/2006 as regards maximum levels for nitrates in foodstuffs.

[90] European Union (2011) Commission Regulation (EU) No 1129/2011 amending Annex II to Regulation (EC) No 1333/2008 of the European Parliament and of the Council by establishing a Union list of food additives.

[91] International Organization for Standardization (1975) ISO 3091:1975 - *Meat and meat products - Determination of nitrate content*.

[92] Mohamed, A. A., Mubarak, A. T., Fawy, K. F., El-Shahat, M. F. (2008)Modification of AOAC Method 973.31 for Determination of Nitrite in Cured Meats. *Journal of AOAC International* 91(4) 820-827.

[93] Pagliano, E., Meija, J., Campanella, B., Onor, M., Iammarino, M., D'Amore, T., Berardi, G., D'Imperio, M., Parente, A., Mihai, O., Mester, Z. (2019) Certification of nitrate in spinach powder reference material SPIN-1 by high-precision isotope dilution GC-MS. *Analytical and Bioanalytical Chemistry,* 1-11

[94] Usman, A. S., Muazu, J., Abba, S. and Ibrahim, M. K. (2015). Nitrate contents in some vegetable leaves in Sokoto Metropolis, Nigeria. *African Journal of Biochemistry Research,* 9(10), 124-129.

[95] Hsu, J., Arcot, J., Lee, N. A. (2009) Nitrate and nitrite quantification from cured meat and vegetables and their estimated dietary intake in Australians. *Food Chemistry,* 115, 334–339.

[96] European Normalization Committee (2005) Standard Reference EN 12014-4:2005. Foodstuffs - Determination of nitrate and/or nitrite content - Part 4: Ion-exchange chromatographic (IC) method for the determination of nitrate and nitrite content of meat products. *CEN-CENELEC Management Centre*, Brussels.

[97] Siu, D. C, Henshall, A. (1998) Ion chromatographic determination of nitrate and nitrite in meat products. *Journal of Chromatography A,* 804(1-2) 157-60.

[98] Dionex Corporation (2007) Product manual for IonPac AS9-HC, in *Dionex Reference Library version,* 31, 14.

[99] Thompson, M., Ellison, S. L. R., Wood, R., (2002) Harmonized guidelines for single laboratory validation of methods of analysis. *Pure and Applied Chemistry,* 74(5):835-855.

[100] European Union (2002) Commission regulation (EC) No 657/2002. *Official Journal of the European Communities,* L221:8–36.

[101] D'Amore, T., Di Taranto, A., Vita, V., Berardi, G., Iammarino, M. (2019) Development and validation of an analytical method for nitrite and nitrate determination in meat products by capillary ion chromatography (CIC). *Food Analytical Methods,* 1-10

BIOGRAPHICAL SKETCH

Marco Iammarino

Italian National Reference Centre for the Detection of Radioactivity in Feed and Foodstuff. Chemistry Department - Istituto Zooprofilattico Sperimentale della Puglia e della Basilicata, Foggia, Italy.

Education:

2006. I.T.I. G.Galilei of Salerno, Italy. Chemical Surveyor Qualification.

2004. University of Foggia, Italy. 1st level Master degree: *"Expert of Fish Food Supply Chain Innovation."*

2001. University of Foggia, Italy. Food Technologist Qualification.

2000. University of Foggia, Italy. Bachelor Degree: Food Science and Technology. Thesis title: *"Production of Canestrato Pugliese mini-cheeses by use of Non Starter Lactic Acid Bacteria (NSLAB)."* Advisor: Prof. M. Gobbetti.

1995. I.T.I.S. S.Altamura of Foggia, Italy. Secondary Technical School: Industrial Chemistry Degree.

Business Address: Istituto Zooprofilattico Sperimentale della Puglia e della Basilicata. Via Manfredonia, 20, 71121 Foggia – Italy

Research and Professional Experience: Food Technologist and Chemical Surveyor at Chemistry Department of Istituto Zooprofilattico Sperimentale della Puglia e della Basilicata, Foggia, Italy, since 01 March 2002. He deals mainly with:

- Food safety and inspection;
- Analytical chemistry applied to the safety of food productions;
- Development of Research Projects and relating results disclosure;
- Quality Management, validation and accreditation of chemical analytical methods.

Research Experience:

2016-2019. Principal Investigator of Project *"Development of innovative analytical methods for meat products safety and quality assurance,"* financed by Italian Ministry of Health (Roma, Italy) and carried out at Istituto Zooprofilattico Sperimentale of Puglia and Basilicata (Foggia, Italy) in collaboration with University of Foggia (Italy), and Istituto Zooprofilattico Sperimentale of Lombardia and Emilia Romagna (Brescia, Italy).

2014-2015. Collaborator of Project "*Dietary exposure to trace elements and radionuclides: the methodology of the Italian Total Diet Study 2012-2014,*" proposed and carried out by Italian Ministry of Health (Roma, Italy), Italian National Health Institute (Roma, Italy), Italian National Institute for Food and Nutrition (Roma, Italy) and Istituti Zooprofilattici Sperimentali of Puglia and Basilicata (Foggia, Italy) and of Umbria and Marche (Perugia, Italy).

2010-2013. Collaborator of Project *"Monitoring program for the research of physical and chemical contaminants in food commodities imported from Bosnia Herzegovina and from Kosovo,"* proposed and carried out by Italian Ministry of Health (Roma, Italy), Italian National Health Institute (Roma, Italy) and Istituti Zooprofilattici Sperimentali of Puglia and Basilicata (Foggia, Italy) and of Umbria and Marche (Perugia, Italy).

2010. Chemistry Department of Istituto Zooprofilattico Sperimentale della Puglia e della Basilicata, Foggia, Italy. "*Optimisation of a confirmatory analytical method for the determination of DON (Deoxynivalenol) and NIV (Nivalenol) in cereals by HPLC.*"

2009-2010. Chemistry Department of Istituto Zooprofilattico Sperimentale della Puglia e della Basilicata, Foggia, Italy. *"Optimization and validation of an analytical method for the determination of Fumonisins FB1 and FB2 in cereals by HPLC with post-column derivatisation and fluorimetric detection: comparison with immunoenzymatic method and monitoring on real samples."*

2008-2009. Chemistry Department of Istituto Zooprofilattico Sperimentale della Puglia e della Basilicata, Foggia, Italy. *"Finalization of an innovative method for the determination of Aflatoxins B1, B2, G1, G2 in foodstuffs by HPLC with photochemical derivatisation."*

2007-2008. Chemistry Department of Istituto Zooprofilattico Sperimentale della Puglia e della Basilicata, Foggia, Italy. *"Environments and breeding technologies characterisation for quality and environmental certification of sea cage-breeding productions from Puglia."*

2007-2008. University of Foggia and BIOAGROMED Centre, Italy. *"Innovative systems of production and packaging for meat products."*

2006-2007. Chemistry Department of Istituto Zooprofilattico Sperimentale della Puglia e della Basilicata, Foggia, Italy. *"Validation of an innovative analytical method for the determination of biogenic amines in food matrices by ion chromatography with conductivity detection."*

2005-2006. Chemistry Department of Istituto Zooprofilattico Sperimentale della Puglia e della Basilicata, Foggia, Italy. *"Monitoring of Aflatoxin M1 in milk and dairy products also in raw materials used for food producing."*

2003-2005. Chemistry Department of Istituto Zooprofilattico Sperimentale della Puglia e della Basilicata, Foggia, Italy. *"Development of innovative methodologies for identification and quantification on biogenic amines in food matrices."*

2000-2002. C.L.A. s.c.r.l., Foggia, Italy. "*Study of technological innovation in green table olives "Picholine," "Bella di Cerignola" and "Sant'Agostino" production."*

2000. University of Foggia, Italy. *"Production of Canestrato Pugliese mini-cheeses by use of Non Starter Lactic Acid Bacteria (NSLAB)."*

Professional Appointments:

Reviewer activity: *Trends in Food Science & Technology; Science of the Total Environment; Journal of Agricultural and Food Chemistry; Food Chemistry; Meat Science; Food and Chemical Toxicology; Food Additives & Contaminants: part A; Journal of Food Composition and Analysis; Journal of Chromatography B; International Journal of Food Science & Technology; Frontiers in Chemistry; Frontiers in Physiology; Journal of Analytical Methods in Chemistry; Frontiers in Microbiology; Algal Research; Environmental Science and Pollution Research; Ecotoxicology and Environmental Safety; International Journal of Environmental Analytical Chemistry; Archives of Agronomy and Soil Science; Journal of Food Processing and Preservation; Journal of Food Quality; Natural Product Research; Journal of Food Measurement and Characterization; Acta Paediatrica; Biological*

Agriculture & Horticulture; Journal of Consumer Protection and Food Safety and other 36 International Journals.

Editorial Board Member of "Frontiers in Nutrition," "Journal of Food Research," "Acta Scientific Nutritional Health" and "Emergent Life Sciences Research."

Guest Editor: "*Recent advances in meat products quality & safety improvements and assurance*" – International Journal of Food Science & Technology (Special Issue), *in progress*.

Lead Guest Editor: "Emerging Issues in Food Safety, Food Additives: Risk Assessment, Analytical Methods and Replacement in Foodstuffs" - *Journal of Food and Nutrition Sciences* (Special Issue) Volume 3, Issue 1-1, 2015.

Honors:

Official H-index (Scopus): 12

2018. Top 1% of reviewers in each field who performed the most verified pre-publication peer reviews on Publons for the 2018 global Peer Review Awards.

2011. XIII National Congress of the Italian Society of Veterinary Laboratory Diagnosticians, Trani (BA), Italy. Recognition for scientific activity for the contribution titled: *"Validation and application of multi-residue analysis of eight anticoagulant rodenticides by high-performance liquid chromatography with fluorimetric detection."*

2003-2010. Scholarship at Istituto Zooprofilattico Sperimentale della Puglia e della Basilicata, Foggia, Italy.

2007-2008/1997-2000. Scholarship at University of Foggia, Italy.

Publications from the Last 3 Years:

D'Amore, T., Di Taranto, A., Vita, V., Berardi, G., Iammarino, M. (2019). "Development and validation of an analytical method for nitrite and nitrate determination in meat products by capillary ion chromatography (CIC)". *Food Analytical Methods,* DOI:10.1007/ s12161-019-01529-0.

Iammarino M., Mentana A., Centonze D., Palermo C., Mangiacotti M., Chiaravalle A.E. (2019). "Chromatographic determination of 12 dyes in meat products by HPLC-UV-Diode array detection." *MethodsX*, 856-861.

Pagliano E., Meija J., Campanella B., Onor M., Iammarino M., D'Amore T., Berardi G., D'Imperio M., Parente A., Mihai O., Mester Z. (2019). Certification of nitrate in spinach powder reference material SPIN-1 by high-precision isotope dilution GC–MS. *Analytical & Bioanalytical Chemistry*, DOI: 10.1007/s00216-019-01803-4.

Iammarino M., Mentana A., Centonze D., Palermo C., Mangiacotti M., Chiaravalle A.E. (2019). "Simultaneous determination of twelve dyes in meat products: development and validation of an analytical method based on hplc-uv-diode array detection." *Food Chemistry*, 285, 1-9.

Iammarino M., dell'Oro D., Bortone N., Mangiacotti M., Chiaravalle A.E. (2018). "Radiostrontium accumulation in animal bones: development of a radiochemical method by ultra low level liquid scintillation counting for its quantification." *Veterinaria Italiana*, 54(1), 41-47.

Iammarino M., Mentana A., Centonze D., Palermo C., Mangiacotti M., Chiaravalle A.E. (2018). "Optimization and validation of an analytical method by HPLC-UV-Diode array detection for the simultaneous determination of 12 dyes in meat products." *Journal of Food, Nutrition and Public Health*, 2, 33.

Iammarino M., Tomaiuolo M., Chiaravalle A.E., Mangiacotti M. (2018). "Sample irradiation coupled to electronic spin resonance analysis for identifying a mechanically separated meat." *Journal of Food, Nutrition and Public Health*, 2, 61.

Iammarino M., Di Taranto A., Centonze D. (2017). "Determination of Sulphiting Agents in Raw and Processed Meat: Comparison Between a Modified Monier-Williams Method and the Direct Analysis by Ion Chromatography with Conductometric Detection." *Food Analytical Methods*, 10(12), 3956–3963;

Carrabs g., Smaldone G., Carosielli L., Girasole M., Iammarino M., Chiaravalle E. (2017). "Detection of sulfites in fresh meat preparation commercialised at retail in Lazio Region." *Italian Journal of Food Safety*, 6:6482, 93-95;

Iammarino M. (2017). "Development of innovative analytical methods for meat products safety and quality assurance "The MPSQA project." *Journal of Food Microbiology, Safety & Hygiene*, 2(2), 28.

Iammarino M. (2017). "Natural levels of nitrites, nitrates, sulphites and ascorbic acid in fresh meat preparations: A contribution to the estimation of maximum admissible limits." *Journal of Food Microbiology, Safety & Hygiene*, 2(2), 29.

Iammarino M., Miedico O., Mangiacotti M., Bortone N., Damiano R., Chiaravalle A.E. (2017). "Assessment of beta-emitting radionuclides (Strontium-90) levels in plant-derived foods marketed in Italy. In: Kathy Schneider Ed., *Advances in Vegetable Consumption and Health Research*. Nova Science Publisher, Hauppauge NY, USA, 55-76.

Iammarino M., Di Taranto A., Ientile A.R. (2017). "Sulphur dioxide in meat products: 3-year control results of an accredited Italian laboratory." *Food Additives & Contaminants: Part B*, 10(2): 99-104;

Iammarino M., Marino R., Albenzio M. (2017). "How meaty? Detection and quantification of adulterants, foreign proteins and food additives in meat products." *International Journal of Food Science & Technology*, 52: 851-863 (*Invited Review*);

Miedico O., Iammarino M., Tarallo M., Chiaravalle A.E. (2017). "Application of inductively coupled plasma mass spectrometry for trace element characterisation of equine meats." *International Journal of Food Properties*, 20(12), 2888-2900;

Iammarino M., dell'Oro D., Bortone N., Mangiacotti M., Damiano R., Chiaravalle A.E. (2016). "Food stuffs radio contamination by ^{90}Sr: analytical methods, mean levels in food and contribution to risk assessment." *Journal of Food Processing & Technology*, 7(12), 69;

Iammarino M., dell'Oro D., Bortone N., Mangiacotti M., Damiano R., Chiaravalle A.E. (2016). "Radiostrontium levels in foodstuffs: 4-years control activity by Italian Reference Centre, as a contribution to risk assessment." *Food Chemistry*, 210, 344-354;

Miedico O., Iammarino M., Paglia G., Tarallo M., Mangiacotti M., Chiaravalle A.E. (2016). "Environmental monitoring of the area surrounding oil wells in Val d'Agri (Italy): element accumulation in bovine and ovine organs." *Environmental Monitoring and Assessment*, 188(6): 1-11;

Iammarino M., Di Taranto A., Ientile A.R. (2016). "Ion chromatography with conductivity detection as accredited technique for quantifying sulphites in meat products and crusteceans." *Industrie Alimentari*, 566, March 2016, 3-11;

Muscarella M., Armentano A., Iammarino M., Mayer M., Amorena M. (2016). "Survey of animals poisoning by anticoagulant rodenticides in the regions of Apulia and Basilicata (Italy)." *Veterinaria Italiana*, 52(2), 153-159;

Iammarino M., dell'Oro D., Bortone N., Mangiacotti M., Chiaravalle A.E. (2016). "Optimisation and Validation of a Multi-matrix Ultrasensible Radiochemical Method for the Determination of Radiostrontium in Solid Foodstuffs by Liquid Scintillation Counting." *Food Analytical Methods*, 9(1), 95-104.

In: Agricultural Research Updates. Vol. 28 ISBN: 978-1-53616-253-0
Editors: P. Gorawala, S. Mandhatri

Chapter 6

ASSESSMENT OF THE NUTRITIONAL POTENTIAL OF *ABELMOSCHUS MOSCHATUS* (L.) MEDIK (MALVACEAE)

Bindu R. Nair **[*]** ***and Manju Manoharan***
Department of Botany, University of Kerala, Kariavattom, Thiruvananthapuram, Kerala, India

ABSTRACT

Biodiversity prospecting is concerned with exploration, collection and screening of natural resources along with associated indigenous knowledge for their beneficial utilization. Fortunately, the Indian subcontinent is gifted with many food crops and genetically diverse populations of their wild relatives. The fruits of many of these wild plants are consumed by the rural folk and literature is replete with reports of their medicinal worth. Also, the fruits of some of these plants are rich sources of nutrients, vitamins, minerals and anti-oxidants. Therefore, it is imperative that such lesser known edible plants should be collected from the wild, documented, studied and conserved. *Abelmoschus moschatus*

[*] Corresponding Author's E-mail: bindunair_r@yahoo.co.in.

(L.) Medik, (Malvaceae), commonly called 'ambrette plant' or 'musk mallow' is a tropical semi-wild, perennial shrub native to India, and a close ally of *A. esculentus*, the common okra. The present study focused on the scientific validation of the medicinal and edible utility of fruits of *A. moschatus* using standard protocols so that it can account for potential commercial utilization. About 15 nutritional and five anti-nutritional constituents of the fruits of *A. moschatus* were tested and compared with that in *A. esculentus*. Just as in common okra, the carbohydrate and crude fibre content was relatively higher than the protein and lipid content. The vitamin content however, was higher than in common okra. The anti-nutrient content was comparable, except oxalates and phytates. The study also included the evaluation of the phytochemical constituents of the fruits. Seven of the eleven phytochemicals tested such as flavonoids, phenols, tannins, glycosides, cardiac glycosides, saponins and coumarins were present in fruits. Based on the results of the present investigation, it is evident that the fruits of *A. moschatus* can be included in the diet just as *A.esculentus* and the plant deserves due promotion to commercial cultivation.

Keywords: *Abelmoschus moschatus,* nutritional and anti-nutritional constituents, phytochemicals

INTRODUCTION

Food crops such as cereals, tubers, fruits and vegetables satisfy the nutritional needs of man and determine the nutritional security of a country. In India, the production of horticultural crops especially fruits and vegetables, far exceeds cereals. India is considered the second largest producer of horticultural crops in the world and over the last five years, vegetable production has been relatively high compared to the other crops. However, more than half of the total produce is contributed by only a few popular vegetables such as potato, onion, tomato, brinjal and okra, while the rest remain underutilized (Vanitha et al., 2013; Anonymous, 2017). Among these, many are wild edibles, found distributed in forests, unknown to people in the urban regions but consumed substantially by rural and tribal people (Mahapatra and Panda, 2012). Ignorance about their edible quality, nutritional worth or potential market value possibly limits their use and therefore they account for only 10% of the total crop produce. However, of late, wild edibles are gaining importance as sources of valuable micronutrients (FAO, 2018). Therefore, against all odds, the genetic resources of such wild edibles should be collected, conserved and maintained.

Okra (*Abelmoschus esculentus* (L.) Moench, Family–Malvaceae), is an important vegetable crop enjoying wide distribution in both the tropical and temperate regions of the world. India ranks first in the production of okra (73%) in the world (Vanitha et al., 2013). This vegetable provides an important input of vitamins and mineral salts, including calcium which are often lacking in the diet of people in the developing countries. The other cultivated species of *Abelmoschus* include, *A. caillei, A. manihot* and *A. moschatus*. Of these, *A.caillei* is cultivated in Africa for its tender fruits, *A. manihot* and *A. moschatus* are chiefly valued for their tender leaves and aromatic seeds respectively (IBPGR, 1991; Lokesh, 2017).

The present chapter deals with nutritional evaluation of *Abelmoschus moschatus* (L.) Medik. *Abelmoschus moschatus*, is an erect annual hirsute herb with palmately compound leaves and yellow flowers. *Abelmoschus moschatus* grows in a range of habitats from marshy areas to forest edges at elevations of upto 450 m (Mishra and Naik, 2000). The plant gives fruits within one month and remains a perennial if left undisturbed while *A.esculentus* ia an annual. It is commonly known as the 'ambrette plant' or 'musk mallow' due to the presence of a volatile oil known as 'ambrette oil'or 'musk oil' in the seeds. Ambrette oil of commerce is extracted and used in perfumery, flavouring, cosmetic and agarbathi industries (Rao et al., 2006). The leaves and seeds are valued medicinally as seeds have aphrodisiac, antispasmodic properties prescribed to cure vomiting, intestinal disorders, urinary discharge, nervous disorders, hysteria and skin diseases.

Briefly, in this chapter we discuss the nutritional and phytochemical status of *A. moschatus* fruits and in comparison with okra.

Materials and Methods

Yield Characteristics

Seeds of *A.esculentus* and *A.moschatus* were sown during wet summer and about 20 seedlings grown to maturity in an experimental plot of Department of Botany, University of Kerala, Kariavattom. The plants were irrigated periodically and added organic manure. Upon flowering, various yield parameters were recorded (Table 1).

Productivity of each species was evaluated on the basis of data obtained from the pooled monthly counts of flowers and fruits on individual plants. Data was recorded for a period of six months and the monthly fruit count was taken as soon as the plant attained reproductive maturity.

Correlation analysis was carried out for each species separately using the SPSS (IBM SPSS Statistics 22) package for the ten yield characters.

Table 1. Yield parameters of *A.esculentus* and *A.moschatus* planted in the experimental plots

Sl.No:	Parameters
1	Plant height at flowering stage (cm)
2	Number of flowers per plant(total)
3	Number of pods per plant(total)
4	Length of the pod (cm)
5	Width of the pod (cm)
6	Number of seeds per pod
7	Pod weight (g)
8	Number of pods/kg
9	Mean weight of hundred seeds(g)
10	Average fruit yield per plant per month

Nutritional Analysis

A pre-specified quantity of the dried fruit powder sample of *A.moschatus* and *A.esculentus* (~100 mg to 4 gm appropriate for each test and proportional to the other additives) was taken for nutritional analysis and the following tests were conducted.

Table 2. Nutritional analysis of *A.esculentus* and *A.moschatus*

Sl. No	Parameters	Method adopted (References)
1	Percentage (%) of moisture	Adetuyi et al., (2011)
2	Crude fibre (%)	*Maynard (1970)
3	Total carbohydrate (g/100g)	*Hedge and Hofreiter (1962)
4	Reducing sugar (g/100g)	*Thayumanavan and Sadasivam, (1984)
5	Crude protein (g/100g)	*Lowry et al., (1951)
6	Total lipid (mg/100g)	*Bligh and Dyer (1959)
7	Energy value (Kcal)	*Asibey-Berko and Tayie (1999)
8	Chlorophyll a (mg/100g)	*Arnon (1949)
9	Chlorophyll b (mg/100g)	*Arnon (1949)
10	Total chlorophyll (mg/100g)	*Arnon (1949)
11	Total carotene(mg/100g)	Thimmaiah (2004)
12	Phenol (mg/100g)	Sadasivam and Manickam (2008)
13	Thiamine (mg/100g)	Sadasivam and Manickam (2008)
14	Riboflavin (mg/100g)	Sadasivam and Manickam (2008)
15	Ascorbic acid (mg/100g)	Sadasivam and Manickam (2008)

* Cited from Sadasivam and Manickam (2008).

Table 3. Mineral Analysis of *A.esculentus* and *A.moschatus*

Parameters	Test methods
Nitrogen (N), Phosphorous (P), Potassium(K), Calcium(Ca)	AOAC, 2012
Magnesium (Mg), Zinc (Zn), Copper (Cu), Iron (Fe), Manganese (Mn)	*ICP-MS Analysis

* ICP-MS Analysis: The inductively coupled plasma mass spectrometry system Thermo Scientific ICAP Qc was used for this work. The external calibration solutions were prepared from standard certified multi element solution (MERCK).

Table 4. Anti-nutritional analysis of *A.esculentus* and *A.moschatus*

Parameters	Method adopted (References)
Oxalates, Saponins Phytic acid, Cyanogenic glycosides	AOAC, 1984
Nitrates	Cataldo et al., 1975

Phytochemical Analysis

About 10 g of fruit powder was extracted for two days by maceration with 100 ml of methanol and distilled water, at room temperature. The extract was filtered and filtrate was collected and used for the analysis of alkaloids, flavonoids, phenols, tannins, phlobatannins, quinones, anthraquinones, glycosides, cardiac glycosides, saponins, coumarins

For qualitative analysis, standard biochemical tests were adopted (Harborne, 1998). Quantitative analysis was carried out for flavonoids, phenols and tannins (Singh et al. 2012).

RESULTS

Table 5. Morphological and yield parameters of *A.moschatus* and *A.esculentus*

Sl.No	Parameters	*A.moschatus*	*A.esculentus*
1	Plant height at flowering stage (cm)	147.80 ± 21.39	161.00 ± 17.81
2	Number of flowers per plant(total)	62.60 ± 13.56	36.30 ± 5.66
3	Number of pods per plant(total)	58.60 ± 12.93	32.60 ± 7.12
4	Length of the pod (cm)	7.80 ± 1.63	14.10 ± 2.69
5	Width of the pod (cm)	11.00 ± 1.83	6.00 ± 1.12
6	Number of seeds per pod	79.20 ± 9.00	77.40 ± 13.44
7	Pod weight(g)	7.70 ± 1.19	14.60 ± 2.12
8	Number of pods/kg	88.80 ± 11.44	47.60 ± 8.19
9	Mean weight of hundred seeds(g)	3.70 ± 0.57	11.72 ± 2.09
10	Average fruit yield per plant per month	9.70 ± 2.06	5.43 ± 1.60

The data on morphological and yield parameters of are recorded in Table 2.

Comparative growth and yield data (Table 2) of *A.moschatus* and *A.esculentus* revealed that plants of *A.esculentus* were taller than *A.moschatus* while the number of flowers and fruits produced was higher in *A.moschatus* compared to *A.esculentus.* The pods were broader, smaller and light-weight in *A.moschatus* compared to *A.esculentus* (Table 5, Figures 1, 2, 3). The average yield per plant per month was greater for *A.moschatus* compared to *A.esculentus* (Figures 4, 5).

Figures 1-3. *Abelmoschus moschatus* 1) Habit 2) Flower 3) Fruit.

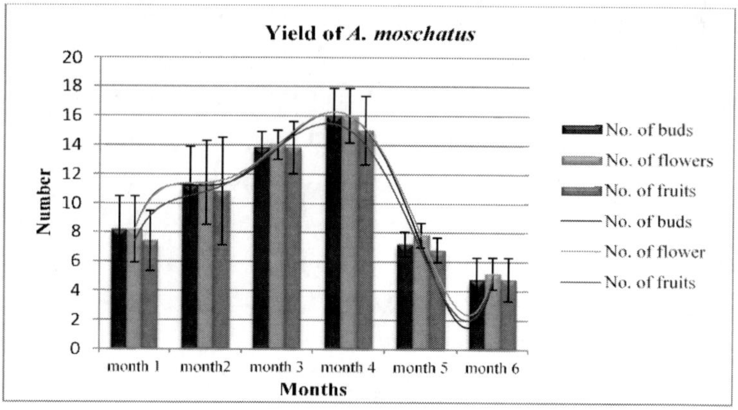

Figure 4. Fruit yield of *A. moschatus* over a period of six months.

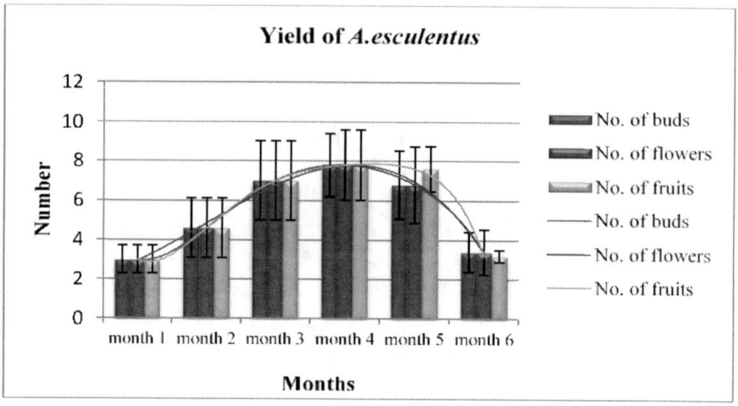

Figure 5. Fruit yield of *A. esculentus* over a period of six months.

Coloured lines represent the trend of the yield.

The results of the correlation co-efficient analysis are given in Tables 6 and 7.

The highest correlation was found between the number of flowers and no. of pods in both the species. Significant positive correlation could be noticed between g, i; a,b; d,f; and negative correlation between b,h; c,h; f,h for *A. moschatus*. Only a few characters showed correlation in *A. esculentus*. Significant positive correlation could be noticed for d,j; i,j and a,b. for *A. esculentus*. The data on nutritional, mineral and anti-nutritional analyses are provided in Tables 8, 9 and 10 respectively.

Table 6. Correlation analysis of ten characters of *Abelmoschus moschatus*

	a	b	c	d	e	f	g	h	i	j
a	1									
b	.762*	1								
c	.590	.956**	1							
d	.479	.589	.544	1						
e	.185	.328	.334	.537	1					
f	.108	.340	.308	.720*	.264	1				
g	.086	.120	.120	.401	.612	.385	1			
h	-.579	-.789**	-.720*	-.529	-.410	-.638*	-.421	1		
i	.015	-.193	-.228	.420	.577	.253	.803**	-.002	1	
j	.382	.250	.317	.146	.325	-.404	-.063	.030	.047	1

*. Correlation is significant at the 0.05 level (2-tailed).

**. Correlation is significant at the 0.01 level (2-tailed).

Table 7. Correlation analysis of ten characters of *Abelmoschus esculentus*

	a	b	c	d	e	f	g	h	i	j
a	1									
b	.635*	1								
c	.582	.883**	1							
d	.295	.203	.008	1						
e	.159	-.170	.222	-.367	1					
f	.124	.618	.355	.614	-.582	1				
g	.122	-.164	.029	.164	.215	-.300	1			
h	.054	-.175	-.041	.371	-.109	.019	.375	1		
i	.005	-.055	-.080	.550	.189	.305	-.171	.086	1	
j	.591	.197	.225	.699*	.251	.169	.237	.363	.679*	1

where, **a**-Plant height, **b**-Total no. of flowers produced per plant, **c**- Total no. of pods produced per plant, **d**-Length of pod, **e**-Width of pod, **f**-No. of seeds per pod, **g**- Pod weight, **h**-No. of pods/kg, **i**-Mean weight of hundred seeds, **j**-Average fruit yield per plant per month.

*. Correlation is significant at the 0.05 level (2-tailed).

**. Correlation is significant at the 0.01 level (2-tailed).

Table 8. Nutritional parameters of *A.moschatus* and *A.esculentus*

Sl. No	Parameters	*A.moschatus*	*A.esculentus*
1	% of moisture	88.56 ± 0.35	86.03 ± 0.47
2	Crude fibre (%)	31.24 ± 0.18	15.11 ± 0.18
3	Total carbohydrate (g/100g)	7.01 ± 0.42	6.84 ± 0.41
4	Reducing sugar (g/100g)	1.33 ± 0.04	1.31 ± 0.04
5	Crude protein (g/100g)	2.31 ± 0.07	2.1 ± 0.26
6	Total lipid (mg/100g)	0.47 ± 0.01	0.12 ± 0.01
7	Energy value (Kcal)	34.62 ± 1.76	30.54 ± 2.13
8	Chlorophyll a (mg/100g)	0.65 ± 0.01	0.40 ± 0.01
9	Chlorophyll b (mg/100g)	0.66 ± 0.01	0.24 ± 0.01
10	Total chlorophyll (mg/100g)	0.64 ± 0.01	0.24 ± 0.01
11	Total carotene(mg/100g)	0.35 ± 0.14	0.27 ± 0.08
12	Phenol (mg/100g)	0.02 ± 0.01	0.01 ± 0.00
13	Thiamine (mg/100g)	0.40 ± 0.13	0.14 ± 0.08
14	Riboflavin (mg/100g)	0.47 ± 0.08	0.06 ± 0.02
15	Ascorbic acid (mg/100g)	36.43 ± 14.06	33.46 ± 12.23

Table 9. Mineral analysis of *A.moschatus* and *A.esculentus*

Sl.No.	Elements	*A. moschatus*	*A. esculentus*
1	N(%)	2.450	2.485
2	P(%)	0.328	0.307
3	K(%)	2.230	2.208
4	Ca (ppm)	295.200	534.600
5	Mg (ppm)	250.100	250.290
6	Zn (ppm)	65.600	44.600
7	Cu (ppm)	29.200	52.700
8	Fe (ppm)	76.500	210.100
9	Mn (ppm)	42.100	44.100

Table 10. Anti-nutritional parameters of *A.moschatus* and *A.esculentus*

Sl. No.	Parameters (mg/g)	*A.moschatus*	*A.esculentus*
1	Oxalates	3.06 ± 0.01	5.36 ± 0.01
2	Saponins	2.53 ± 0.01	2.09 ± 0.01
3	Phytic acid	1.58 ± 0.01	0.60 ± 0.01
4	Cyanogenic glycosides	0.91 ± 0.01	0.95 ± 0.01
5	Nitrates	0.77 ± 0.01	0.74 ± 0.01

The nutrient content in both the species was comparable. The crude fibre content and carbohydrate content was relatively higher in both species, more so in *A. moschatus*, while protein and lipid content was considerably less. The energy value was comparable but higher for *A. moschatus*. The thiamine and riboflavin contents were higher for *A. moschatus*, while the carotene and ascorbic acid contents were comparable in both the species.

Calcium and Magnesium were relatively high in both the species while Calcium content was higher in *A. esculentus*. Nitrogen, phosphorous and potassium values were comparable for both the species. Zinc was higher in *A. moschatus* but the other minor elements were higher in *A. esculentus.*

The anti-nutritional factors were relatively low in both the samples but the quantity of oxalates was higher in *A. esculentus*, and phytic acid was higher in *A. moschatus.*

Phytochemical Constituents

Table 11. Qualitative phytochemical analysis of extracts of *A.moschatus*

Sl.No.	Parameters	*moschatus*	
		Methanolic extract	**Aqueous extract**
1	Alkaloid	-	-
2	Flavonoid	+	+
3	Phenols	+	+
4	Tannin	+	+
5	Phlobatannins	-	-
6	Quinones	-	-
7	Anthraquinones	-	-
8	Glycosides	+	+
9	Cardiac Glycosides	+	+
10	Saponins	+	+
11	Coumarins	+	+

'+' indicates presence and '-' indicates absence.

Table 12. Quantitative phytochemical analysis of extracts of *A.moschatus*

Sl.No.	Group of phytochemicals	*A.moschatus*	
		Methanolic extract	Aqueous extract
1	Flavonoids (mg QE/g)	84.94 ± 0.962	46.98 ± 1.990
2	Phenols(mg GA E/100g)	50.58 ± 2.667	76.41 ± 2.739
3	Tannins (mg TA E/g)	19.13 ± 1.786	30.57 ± 0.679

The data for qualitative and quantitative phytochemical analyses of the methanol and aqueous extracts are provided in Tables 11 and 12 respectively.

DISCUSSION

Yield Characteristics

The morphological and yield characters of *A.esculentus* and *A.moschatus* showed variations. Plant height was greater for *A.esculentus* but the number of flowers and consequently the fruits produced were higher in *A.moschatus*. The pods were broader, smaller and light-weight in *A.moschatus*.

The total number of flowers and pods produced per plant varied between the species over a period of six months of observation. The productivity of *A. esculentus* and *A.moschatus* in terms of formation of buds, flowers and fruits gradually increased, attained maximum at 3rd-4th months and thereafter declined (Figures 4 and 5). *Abelmoschus moschatus* had a higher value for both the mean number of flowers and pods produced per plant.

The tender green pod is considered as the most important and economical part of the plant since it is utilized as a vegetable throughout the world. Consequently, fruit length in consonance with pod number and pod weight are most important determinants of production or yield. The present study observed that there was a significant variation in the mean pod length, width and pod weight between the two species under investigation. The pods of *A.esculentus* were longer than broad and in *A.moschatus* broader than long. Pod weight for *A.esculentus* was greater compared to *A.moschatus.* The two species revealed a noteworthy difference in the number of pods per kilogram. *Abelmoschus esculentus* exhibited a greater mean number of pods/kg than that of *A.moschatus*. Mean weight of hundred seeds was recorded high in *A.moschatus*. Mean weight of hundred seeds is usually associated with pod yield and other yield attributes in a seed bearing plant.

The relationship between the two species based on pairs of morphological and yield characters was studied. The correlation analysis revealed that plant height and number of flowers, number of flowers and number of pods were significantly positively correlated in both the species. The present study revealed that plant height and total number of flowers was positively correlated; thereby plant height influenced the yield. Plant height at maturity is of particular interest for breeding programmes. (Akinyele et al., 2006). *Abelmoschus moschatus* attained reproductive maturity earlier characterized by early flowering. Higher yield was also observed in *A. moschatus*.

Nutritional Factors

Analysis of moisture content in the fruit samples of the two species of *Abelmoschus* under the present study showed that the values are almost comparable in both *A.moschatus* (88.56 ± 0.35%) and *A.esculentus* (86.03 ± 0.47). Similar values have been reported earlier (Adetuyi et al., 2011).

Fresh okra pods are the most important vegetable source of viscous fiber. Dietary fibers are important components in the diet as it slows down sugar assimilation through and lowers cholesterol (Kendall & Jenkins, 2004, Ngoc et al., 2008). The present study showed that *A.moschatus* fresh fruits contained double the amount of crude fibre than *A.esculentus*. The fresh six to seven days old okra pod used in the study contained high amount of fibre and other nutrients. This agrees with the findings of Kumar et al. (2016), who noticed that seven days old fresh okra pods had the highest concentration of nutrients. High fibre content of *A.moschatus* suggests that it can be suitably included in the diet but should be consumed earlier than *A.esculentus* as it tends to become fibrous. There was no significant difference in the amounts of carbohydrate and reducing sugar in *A.esculentus* and *A.moschatus*, while the protein and lipid content was relatively low. Diet is nutritionally satisfactory, if it contains high calorific value (Nwofia et al., 2012). *Abelmoschus moschatus* had a greater energy value than *A.esculentus*.

The chlorophyll content was higher for *A.moschatus.* Since green vegetables contain the chlorophyll constituent, chlorophyllin characterized with enormous health benefits, consumption of *A. moschatus* may be considered suitable (Ebermann et al., 1996). The fruits of *A.moschatus* contained higher amount of total carotene than that of *A.esculentus.* Vitamins are essential for proper body development. Thiamine (vitamin B1), and Riboflavin (vitamin B2) are important for body growth, red blood cell production and releasing energy from carbohydrates. Ascorbic acid or vitamin C is an essential anti-oxidant that helps in the growth and repair of body tissues. Thiamine and riboflavin contents were higher for *A. moschatus*, while the carotene and ascorbic acid contents were comparable in both species. Rai et al., (2012) had reported the presence of Vitamins B1, B3, B5, B6, B9 and K in *A. esculentus* highlighting its nutraceutical potential.

Minerals such as Calcium and Magnesium were relatively high in both the species while Calcium content was higher in *A. esculentus*. Nitrogen, phosphorous and potassium values were comparable for both the species. Zinc was higher in *A. moschatus* but the other minor elements were higher in *A. esculentus*

Anti-Nutritional Factors

Anti-nutrients are natural or synthetic compounds that interfere with the absorption of nutrients. The anti-nutritional factors analyzed in the present study, were oxalates, saponins, phytic acid, cyanogenic glycosides and nitrates. Oxalates usually present in members of spinach family cause their effect by binding to calcium and preventing its absorption in the human body. The present study revealed that there was a slight increase in the quantity of oxalates in the pods of *A.esculentus* compared to *A.moschatus*. Phytic acid has a strong binding affinity to minerals such as calcium, magnesium, iron, copper and zinc, making them unavailable for absorption in the intestine (Adetuyi et al., 2011). There was a slight increase in the quantity of phytic acid in the pods of *A.moschatus* compared to *A.esculentus*. Saponins, cyanogenic glycosides and nitrates were relatively less in both samples. The anti-nutrient content was relatively low in both the samples and is therefore suitable for human consumption

Phytochemicals

Abelmoschus esculentus is reported to possess pharmacological properties like antidiabetic, antioxidant, nootropic, eye, heart disease and neurological disorders (Sindhu and Puri, 2016; Hunag et al., 2017). The therapeutic properties of plant materials are usually attributed to the presence of phytochemicals, especially the secondary metabolites in plants. Fresh samples as well as extracts are used to screen for the presence of phytochemicals. The methanol and aqueous extracts of *A. moschatus* gave positive results for seven of the eleven phytochemicals tested. Depending upon the results of the preliminary qualitative analyses, the present study was mainly concentrated on the quantification of phenols and polyphenolic compounds such as flavonoids and tannins in the plant samples. Flavonoid content was higher in the methanolic extract while phenols and tannins in the aqueous extracts of *A. moschatus*. Previous reports showed that flavonoids, tannins and phenols tested positive and flavonoid content was found to be greater in comparison to the phenol and tannin contents, in the methanolic and aqueous extracts of *A. esculentus* (Kumar, 2014, Bogninou et al., 2018).

The nutritional and pharmaceutical potential of *A. esculentus* has been discussed earlier in many reviews (Lokesh, 2017). The results of the present study on the nutritional and phytochemical profile of *A.moschatus* qualify it for an equal standing with *A.esculentus*. As discussed above,

A.moschatus has comparably higher values for carbohydrates, crude fiber, essential vitamins, minerals, flavonoids and considerably lower amounts of anti-nutritional factors (ANFs), phenols and tannins suggestive of its importance in human health and the possibility of utilizing it for the development of nutraceutical formulations.

REFERENCES

Adetuyi F.O., Osagie A.U. and A.T. Adekunle A.T. "Nutrient, antinutrient, mineral and zinc bioavailability of okra *Abelmoschus esculentus* (L) Moench. Variety." *American Journal of food and nutrition* 1(2) (2011): 49-54, doi:10.5251/ajfn.2011.1.2.49.54 © 2011, ScienceHuβ, http://www.scihub.org/AJFN.

Akinyele, B.O., Oseikita, O.S. "Correlation and path coefficient analyses of seed yield attributes in okra (*Abelmoschus esculentus* (L.)Moench)."*African Journal of Biotechnology* 5(2006):1330-1336.

Anonymous. Horticultural Statistics at a Glance 2017, Horticulture Statistics Division, Department of Agriculture, Cooperation & Farmers Welfare, Ministry of Agriculture & Farmers Welfare, Government of India (2017).

AOAC, Association of Official Analytical Chemists. *Official Methods of Analysis*. 14th Edn. Washington DC. (1984).

AOAC, Association of Official Analytical Chemists. *Official Methods of Analysis*, AOAC, 19th Edn.United States (2012).

Bogninou G. Sophie Reine, Bigo Agadaba Prudence Houndékon, Gnanwe Marcelline, Agbangnan Dossa C. Pascal, Chabi Nicodème Worou, Yedomonhan Hounankpon, and Avlessi Félicien,. "Phytochemical Composition and Antioxidant Capacity of *Abelmoschus esculentus* l. Fresh Immature Fruits." *American Journal of Food Science and Technology*. 6(5) (2018): 223-227. doi: 10.12691/ajfst-6-5-6.

Cataldo D.A., Harcon, M., Schrader, L.E., Youngs, V.L. "Rapid colorimetric determination of nitrate in plant tissue by nitration of salicylic acid." *Communications in Soil Science and Plant analysis.* 6(1975): 71-80.

Ebermann, R., Alth, G., Kreitner, M., & Kubin, A. "Natural products derived from plants as potential drugs for the photodynamic destruction of tumor cells." *Journal of Photochemistry and Photobiology* B: Biology 36(1996):95-97.

FAO (Food and Agricultural Organization) of the United Nations, Food Outlook, *Biannual Report on Global markets,* (2018). www.fao.org/ publications.

Food and Agricultural Organization of United nation (FAO), *Economic and social development department:* The Statistical division, (2014).http://faostat.fao.org/.

Harborne, J.B. *Phytochemical Methods-A guide to modern techniques of plant analysis.* 3rd Edn (1998). Chapman and Hall Publishers, London.

Huang C.N., Wang C.J., Lin C.L., Lin H.T., and Peng C.H. (2017). The nutraceutical benefits of subfractions of *Abelmoschus esculentus* in treatingtype 2 diabetes mellitus. *PLoS ONE* 12(12) (2017):e0189065. https://doi.org/10.1371/journal.

IBPGR. Report of an international workshop on okra genetic resources, held at the National Bureau for Plant Genetic Resources (NBPGR), New Delhi, India, 8–12 October 1990. International Crop Network Series 5. *International Board for Plant Genetic Resources (IBPGR),* Rome, Italy. p. 133 (1991).

Kendall, C.W.C., Jenkins, D.J.A. A dietary portfolio: maximal reduction of low-density lipoprotein cholesterol with diet. *Current Atherosclerosis Reports* 6 (2004):492-498.

Kumar, R. Yadav, R.K., Bhardwaj, R., Baranwal, V.K., Chaudhary, H. and Vinod. "Assessment of genetic diversity among Okra (*Abelmoschus esculentus*) genotypes for quality traits." *Indian Journal of Agricultural Sciences* 86 (6) (2016): 785-791.

Kumar, S. "Physicochemical, Phytochemical and toxicity studies on gum and mucilage from plant *Abelmoschus esculentus.*" *The Journal of Phytopharmacology*. 3(3) (2014): 200-203.

Lokesh. "Nutritional and Pharmaceutical Potentials of Okra (*Abelmoschus esculentus*) Plant and Its Biotic Stresses -An Overview." *International Journal of Pure and Applied Bioscience* 5 (4)(2017): 1890-1907 DOI: http://dx.doi.org/10.18782/2320-7051.5197.

Mahapatra, A.K., Panda, P.C. *Wild Edible Fruit Plants of Eastern India.* Regional Plant resources Centre, Bhubaneswar, India, (2012).

Mishra D.K. Naik, S.N. "Cultivation and processing of *Abelmoschusmoschatus*." *Journal of Medicinal and Aromatic Plant Sciences*. 22 (1B) (2000):624-628.

Ngoc, T., Ngo, N., Van, T., Phung, V. "Hypolipidemic effect of extracts from *Abelmoschus esculentus* L. (Malvaceae) on Tyloxapol-induced hyperlipidemia in mice." *Mahidol University Journal of Pharmaceutical Sciences* 35 (1-4) (2008):42-46.

Nwofia, G.E., Nwogu, V.N., Nwofia, B.K. "Nutritional Variation in Fruits and Seeds of Pumpkins (*Cucurbita* spp.) Accessions from Nigeria." *Pakistan Journal of Nutrition* 11(2012):946-956.

Singh, R, Verma, P.K., Singh, G. "Total phenolics, flavonoids and tannin contents in different extracts of *Artemisia absinthium." Journal of Intercultural Ethnopharmacology.* 1(2) (2012): 101-104.

Rao B.R., Rajput D.K. Sastri K.P. Kothari. S.K. Singh C.P and Bhattacharya A.K. "Nutrient and intercropping studies in Ambrette (*Abelmoschus moschatus*)" *Journal of Medicinal and Aromatic Plant Sciences* 28 (2006): 372-376.

Sadasivam S. and Manickam A. *Biochemical Methods*. 3rdEdn. New Age International Publishers, New Delhi (2008).

Rai, S.K., Arora, N, Pandey, N, Meena, R.P., Shah, K and Rai, S.P. "Nutraceutical Enriched Vegetables: Molecular Approaches for Crop Improvement." *International Journal of Pharma and Bio Sciences* 3(2 (2012): 363-379.

Sindhu R.K., Puri, V. "Phytochemical, Nutritional and Pharmacological evidences for *Abelmoschus esculentus* (L.)." *The Journal of Phytopharmacology*. 5(6) (2016):238-241.

Thimmaiah S.K. *Standard Methods of Biochemical Analysis.* Kalyani Publishers, New Delhi (2004).

Vanitha, S.M., Chaurasia, S.N.S., Singh P.M. and Prakash S. Naik. *Vegetable Statistics.* Technical Bulletin No. 51, IIVR, Varanasi 2013.

In: Agricultural Research Updates. Vol. 28 ISBN: 978-1-53616-253-0
Editors: P. Gorawala, S. Mandhatri

Chapter 7

CONSERVATION OF CATTLE GENETIC RESOURCES: CURRENT GLOBAL STATUS AND ADVANCEMENTS IN CRYOBIOLOGY

Andrew W. Taylor-Robinson[1,*], Van Huong Do[2] and James E. Kinder[3]

[1]School of Health, Medical and Applied Sciences,
Central Queensland University, Brisbane, Australia
[2]National Key Laboratory of Animal Cell Technology,
National Institute of Animal Sciences, Hanoi, Vietnam
[3]Department of Animal Sciences,
The Ohio State University, Columbus, OH, US

ABSTRACT

Domestic animal genetic resources, including those of cattle, are important for food security, economic development and the cultural heritage of all countries. As with other livestock and poultry species, many cattle breeds are at risk of extinction. The principal underlying

---

* Corresponding Author's E-mail: a.taylor-robinson@cqu.edu.au

reason for the loss of both beef and dairy breeds is an escalating rate and scale of economic and urban development as a result of the increasing global human population. A direct effect of these associated factors is the great emphasis that is placed on agricultural sustainability and animal production systems to provide sufficient food for the ever-increasing number of humans even though there are decreasing natural resources for this food production. Government policies for cattle conservation have been developed on all continents but there are vast differences in detailed aspects of these policies among countries. Initiatives to conserve cattle genetics, including the provision of gene banks for cattle, are well established in industrialized nations. In contrast, less economically developed countries are currently experiencing problems in safeguarding the preservation of local cattle breeds although there exists an immense diversity of animal genetic resources. This disparity is due to differences in a range of key factors such as national strategies, economic status, engagement of private sector companies and available technological capacity. The successful oversight and management of cattle genetic resources has major implications for the biotechnology industry and thus for the growth of a nation's economy. The combination of applying cryobiology to germplasm conservation and implementing national strategies to maintain living genetic resources is, therefore, essential to ensure the sustainability and diversity of cattle breeds. To this end, the advancement of cryopreservation technologies and the international prioritization of conservation breeding are both imperative. In this chapter, we describe the current global status of conservation of cattle genetic resources and the cryo-technologies that are available to facilitate protection of this valuable natural reservoir of bovine genetic diversity. Worldwide adoption of such a policy has led to unique adaptations by cattle to a great variety of local environments.

Keywords: cattle, conservation, cryobiology, genetic, production

INTRODUCTION

The effective utilization of animal genetic resources is an important aspect in meeting the demands for products of livestock (Paiva, McManus and Blackburn, 2016). Since future requirements for specific genetic attributes are unknown the goal of long-term conservation of genetic resources is fulfilled best by the maintenance of genetic diversity (Taberlet et al., 2011). Considerable changes in the use of animal and plant genetic resources have occurred with intensification and globalization of agriculture. This has been a primary contributing factor in the reduction of regionally developed breeds of animals and types of plants with the result being many of these becoming classified as rare or endangered (Tienhaara, Ahtiainen and Pouta, 2015). In addition, there is a worldwide decrease in diversity of animal genetic sources, and among livestock species, cattle have the rate of greatest loss; for example, 184 indigenous cattle breeds are known to have become extinct (FAO, 2014; FAO, 2015).

A principal reason for the loss of cattle breeds that were developed for food production in particular regions of the world is the reliance on the importation of semen/embryos or animals from bulls of a few internationally recognized breeds. The animals from these highly selected breeds are extremely productive in controlled environments where feed resources are in plentiful supply. As a result of the well-defined genetic merit of animals for meat or milk production, these breeds have become very popular with both farmers and consumers. As a consequence, many of the regional breeds are no longer sought for their produce and thus there has been a marked decline or complete loss of animals of these breeds (Taberlet et al., 2011). The cattle of breeds for which there is international availability of cattle genetics and which have been intensively selected for either meat or milk production have in nearly all countries in which they are found displaced regional multipurpose cattle breeds as preferred sources of meat or milk. This has resulted in a reduction of the traditional cattle breeds that are relatively better adapted to a range of regional environmental conditions to which they have been exposed for many generations (Zander et al., 2013).

A diversity of animal genetic resources will likely be highly beneficial when addressing the future consequences of escalating climate change, particularly considering that most of the internationally popular cattle breeds for beef and milk production have been developed in temperate environments where feed resources are relatively plentiful (Leroy et al., 2018). Therefore, it is important to conserve animal genetic resources, particularly when these have been developed in regional areas where adaptation of the animals and selection of those that are more productive has resulted in regional breeds having some unique genomic qualities that will likely be of value to addressing future needs of animals to be well adapted to regional conditions. This is a major consideration if future meat and milk production is to be sustainable under varying climatic conditions.

Generally, there are two methods (*in situ* and *ex situ*) by which animal genetic resources are conserved (Tamou et al., 2018). Paiva, McManus and Blackburn (2016) characterized *in situ* conservation as the conservation of animals in private, university and non-government organizations. *Ex situ* genetic conservation occurs as a result of the development of gene banks, where animal genetic materials such as germplasm and tissues are stored, and is based on governmental control of preservation of these resources.

Cryopreservation of gametes and embryos allows for storage of genetic materials, and represents an integral and important part of routine practices to utilize assisted reproductive technologies (Leibo, 2008). It is important that there be implementation of cattle semen and embryo recovery programs using cryopreservation of gametes. This approach allows for restoration of breeds if the need should arise for so doing (Lauvie et al., 2008).

In addition, cryopreservation capacity provides for a foundational approach to development of gene banks for the storage of animal genetic resources. Gene banking processes encompass information systems, quantification of genetic diversification and cryopreservation methods (Paiva, McManus and Blackburn, 2016). The field of cryopreservation is an area of active research that has advanced over several decades, resulting in the development of various cryopreservation protocols which have met with varying degrees of success (Do et al., 2019).

This chapter reviews the global conservation status and efforts to protect indigenous cattle breeds from loss of genetic diversity and/or extinction. Furthermore, innovations in cryopreservation systems for storage of the genetic materials of cattle are also addressed.

Global Conservation of Cattle Genetic Resources

Those in agricultural science and livestock production leadership roles of many nations have an increased awareness of conservation of cattle genetics. Some discipline-specific experts consider that the sustainability of native cattle genetic resources is critically important, of 'national treasure' status in the context of a country's food security and sustainable farming. Moreover, when the general public gains awareness of the importance of maintaining genetic diversity of livestock there is strong advocacy for action to be taken to conserve animal genetic resources (Zander, Drucker and Holm-Müller, 2009). It is, however, evident that the conservation status of these genetic resources differs vastly among nations. In many developing countries, the importation of well-established exotic cattle breeds is a commonly occurring practice for the purpose of boosting cattle production. Consequently, this action leads to degradation of indigenous cattle breeds. For example, in Vietnam the use of semen derived from other countries for breeds such as Belgian Blue and Droughtmaster to inseminate local yellow cattle causes a dilution and diminution of the genetic base from this regionally adapted breed for future use in production of meat and milk.

Protection of native cattle breeds is therefore needed and *in situ* conservation of indigenous breeds is being widely implemented. There are many ways of protecting traditional cattle breeds as a genetic resource by ensuring their maintenance in the national reservoirs or within local communities. In East African countries, notably Ethiopia and Kenya, there are focused national efforts to maintain the native breeds of cattle. The decision to preserve these breeds was based on the physiological capacity of these animals to withstand the harsh conditions that exist in the regions where each breed was developed, as well as being vigilant to maintain the cultural identity of the local cattle (large body size and horn characteristics) (Zander and Drucker, 2008). Furthermore, *in situ* conservation of these cattle breeds is a relatively inexpensive approach to protecting this genetic resource (Zander and Drucker, 2008).

Government expenditure on the maintenance of cattle genetic diversification is an important aspect of a nation's capacity to sustain genetic diversification of cattle. For example, the investments for local farmers to maintain Borana cattle in Ethiopia and Kenya instead of converting to use of cattle from other countries that are noted for meat or milk productivity has proved an effective approach to preserving the Borana breed as a genetic resource which can be drawn upon if the need should arise in future (Zander, Drucker and Holm-Müller, 2009).

The Food and Agriculture Organization of the United Nations has recommended that the conservation of breeds should focus on live animals (FAO, 2013). In some countries the development of gene banks has been considered as an alternative to maintaining live animals for sustaining genetically diverse resource populations. For example, in 2016 China established the largest national gene bank in the world, which consists of over 300 million genetic specimens (Ochert, 2016).

In contrast, in developed Western nations in which the funding for conservation is more abundant assisted reproductive technologies are widely utilized, including both *in situ* and *ex situ* conservation methods. Blackburn (2009) has emphasized that cattle germplasm conservation is applied extensively due to the extent to which assisted reproductive technologies is utilized. The use of embryo transfer in cattle, through conducting either superovulation or *in vitro* fertilization techniques, is labour intensive and economically expensive. Both considerations are impediments to the unrestricted adoption of these technologies for germplasm conservation – of particular relevance to less affluent developing countries.

In Europe, the number of animals in many indigenous cattle breeds is reduced compared to historical figures (FAO, 2014). For example, there are marked decreases in population of the native Central European cattle breeds, such as the Polish White-Backed, Lithuanian White-Backed, Ukrainian Grey and Slovak Pinzgauer, which threatens the local, if not global, extinction of these breeds (Kasprszak-Filipeck et al., 2018).

In parallel with *ex situ* conservation, one of the main conservation strategies employed in industrialized nations where there are greater economic resources is the provision of direct support payments for the compensation of farmers to maintain regionally developed genetic resources for the public good (Martin-Collado et al., 2014). As a positive outcome of this subsidization policy, there has been a resurgence in the numbers of the Alistana-Sanabresa cattle breed in Spain (Martin-Collado et al., 2014). As mentioned previously, in developing countries the compensation fees to farmers/ranchers for maintaining regional genetic cattle resources are an important aspect in preserving indigenous local cattle breeds. In developed countries, however, citizens in the private sector play an active role in safeguarding the germplasm of native breeds (Leroy et al., 2018).

The US Government has developed the National Animal Germplasm Program with the purpose of maintaining a diverse genetic base for livestock (Blackburn, 2009). The banking of the genetics of animals from established cattle breeds, such as the Holstein-Friesian in France, the Netherlands and the USA, facilitates conservation of the present genetic base for these breeds. This is important because the genetic diversity of these animals is lessened by intensified selection for productivity, as well as by their use on an international scale as a direct result of the capacity to store and transport gametes to any location in the world. The use of genetic resource banks, therefore, is an essential aspect in providing a more diverse genetic pool when there is need for incorporation of these genetics in future breeding programs (Danchin-Burge, Hiemstra and Blackburn, 2011). It is worth noting that gene banking of animal genetic resources is more cost-effective than the long-term maintenance of herds of animals to serve as a genetic resource pool (Paiva, McManus and Blackburn, 2016).

When this information is considered as a whole, there is clearly a vast difference among countries, primarily because of the varied capacity for capital investments, in advancement of technologies for preserving valuable genetic cattle resources. Furthermore, the conservation of cattle genetic resources must be feasible with regard to technical capacity, which depends on national strategies for conserving indigenous cattle resources for the purposes of food security and sustainable agricultural practices.

Cryopreservation of Cattle Genetic Resources

The cutting-edge cryopreservation methods for gametes and embryos can be used effectively to enable the establishment of gene banks for cattle genetic materials globally. The procedures for cryopreservation of bull semen have been optimized for storage of sperm, as a result of which artificial insemination using frozen bull semen is a commonly performed practice worldwide.

Slow Freezing and Vitrification

The cryopreservation of cattle gametes and embryos provides a means to maintain cattle genetic resources. Furthermore, the establishment of commercial companies that transport cattle embryos worldwide has become a thriving enterprise for the cryopreservation of genetic resources. There are two types of cryopreservation techniques: equilibration and non-equilibration (Leibo, 1989). The former is regarded as a conventional method that uses computerized programmable freezing instruments. Embryos are unavoidably damaged upon being subjected to this slow freezing protocol. When embryonic cells are placed in vessels with subzero temperatures the most frequent damage occurs to the cell membrane and organelles as a consequence of intracellular ice crystal formation and the so-called 'solution effect'. This physical injury can lead to cell death and ultimately cause the embryo to die (Dobrinsky, 2001; Yu et al., 2010).

The alternative procedure of vitrification is defined as the solidification of the entire water-based solution free from intracellular ice crystal formation (Taylor-Robinson et al., 2014). Leibo (2008) concluded that cell survival occurs when the occurrence of intracellular ice crystallization is minimal. Loutradi et al. (2008) reported that during vitrification, cells and the surrounding solution solidify in a glass-like state that prevents ice crystal formation. Moreover, the cooling rate for vitrification, as rapid as 15,000 to 30,000ºC/min, is much greater than for conventional freezing, at 0.3 to 0.5ºC/min (Mahmoudzadeh et al., 1994).

van Wagtendonk-de Leeuw et al. (1997) reported that when transferring *in vivo*-derived cattle embryos that had been cooled by vitrification using a 'VS3a' solution containing 6.5 M glycerol and 6% bovine serum albumin (BSA), in field trials the pregnancy rates were similar to those achieved by transferring frozen *in vivo*-derived cattle embryos. However, to date vitrification has not been used widely in commercial settings. Furthermore, while the vitrification technique is considered as an alternative means of embryo cryopreservation, this method still has limitations compared to the conventional slow freezing procedure. From reviewing the literature in which cryopreservation procedures were assessed, Vajta (2000) concluded that there was a justifiable reason why commercial embryo transfer practices favoured less use of vitrification embryo preservation procedures. It was concluded that with vitrification the formation of ice crystals within embryonic cells is suppressed, whereas conventional slow-rate programmable freezing allows for optimization of conditions so at to reduce the adverse effects of cryoprotectant (CPA) toxicity and chilling injury. Hence, despite its drawbacks conventional slow freezing has become the gold standard for the cryopreservation of *in vivo*-derived cattle embryos (Leibo, 2008).

Cryopreserved cattle embryos, especially those derived by *in vitro* fertilization, have a greater viability after use of vitrification procedures for their cryopreservation. For example, following vitrification of *in vitro*-derived cattle embryos, the viability of embryos is greater upon thawing (Matsunari et al., 2012; Sanches et al., 2013; Do et al., 2018). Moreover, this technology has been utilized not only for cryopreservation of cattle embryos (Palasz et al., 1997; Rios et al., 2010; Sanches et al., 2013; Do et al., 2018), but also for cryopreservation of tissues in a wide range of mammalian species. Examples of this use are: cattle oocytes (Asada et al., 2002; Vieira et al., 2002; Morató et al., 2008); swine embryos (Dobrinsky, 2001; Somfai et al., 2006; Cuello et al., 2007); goat embryos (Yacoub,

Gauly and Holtz, 2010); sheep oocytes (Berlinguer et al., 2007; Shirazi et al., 2010); rabbit embryos (Hochi et al., 2004; Hochi et al., 2010; Lin et al., 2011); horse oocytes (Nowak et al., 2014); human embryos (Loutradi et al., 2008); and human corneal endothelial cells (Fan et al., 2009).

The efficacy of the vitrification procedure for the cryopreservation of oocytes and embryos is the result of considerable innovation in vitrification methods, as described in the following sections.

Basic Principles of Cryopreservation

The protective actions of CPAs are critical to the induction of vitrification and the prevention of intracellular ice crystal formation. Two types of compound are used in freezing solutions, permeable and non-permeable CPAs. The common penetrating agents that are efficacious for cryopreserving ova and embryos are propylene glycol, dimethyl sulfoxide (DMSO), glycerol and ethylene glycol (EG). Each of these compounds has a low molecular weight – propylene glycol (MW: 76.1), DMSO (MW: 78.2), glycerol (MW: 91.1) and EG (MW: 62.1) (Leibo, 2008). Of these compounds, the permeability of EG is greatest, which allows for dilution of CPAs in the open pull straw (OPS) and then direct transfer of the embryos to recipients (Vajta et al., 1996). Leibo (2008) reported that the CPAs penetrate cells at different rates and that this also depends on the developmental stage of the embryo being treated. Consistently, the viability at the morula stage is less than that at the blastocyst stage of embryo development (Mahmoudzadeh et al., 1994; Vajta et al., 1996; Pugh, Tervit and Niemann, 2000; Campos-Chillòn et al., 2006; Yacoub, Gauly and Holtz, 2010). Nonetheless, Ishimori et al. (1993) reported that when a combination of DMSO and EG was used similar results were attained for cryopreservation of both cattle morula and blastocyst when using the same freezing/thawing protocol. As described previously, these chemicals perform crucial functions in protecting embryos by preventing intracellular ice crystal formation. Furthermore, the relatively greater permeability of DMSO and EG prevents the extensive osmotic stress that may be observed when using other penetrating agents (Mucci et al., 2006).

In addition to supplementation with penetrating cryoprotectants for vitrification, non-permeable CPAs are also utilized. Typically, for the purposes of extracellular cryoprotection these are high molecular weight polymers (Checura and Seidel, 2007). For example, when supplementing cryoprotectants with trehalose Berlinguer et al. (2007) noted that there was stabilization of the cell membrane during vitrification. Vajta et al. (1996) reported that the use of sucrose prevented excessive swelling of embryos during thawing. In addition, Asada et al. (2002) concluded that there are benefits to embryo viability with use of polyvinyl alcohol as a substitute for serum in a vitrification solution.

Palasz et al. (1996) reported that inclusion of CPAs in the vitrification solution prevented cell damage due to osmotic replacement of intracellular water before cooling. However, the concentration of CPAs in the vitrification solution is very high. Although the relatively greater level of CPAs is necessary to induce vitrification of a cell suspension (Checura and Seidel, 2007), this leads to toxicity and osmotic injury of embryonic cells (Somfai et al., 2006; Loutradi et al., 2008). There is therefore a combination of different CPAs used in freezing solutions rather than reliance on just one (Vajta et al., 1996). Rapid cooling and warming rates also minimize the toxicity of CPAs (Begin et al., 2003; Seki, Jin and Mazur, 2014).

It is important to note that while CPAs continue to be used in vitrification solutions over time the concentrations used have been gradually reduced. This has led to less toxic effects of CPAs on cattle embryos. Seki, Jin and Mazur (2014) showed that the concentration of CPAs in vitrification solutions can be reduced to half those traditionally used without any change in the viability of *in vitro*-derived embryos. Furthermore, Do et al. (2018) recently reported that vitrification may not be toxic to cattle embryos since they demonstrated no loss during early pregnancy in recipients receiving vitrified embryos.

Recent Attention on the Role of Warming Rates

Cryobiologists have verified the importance of warming rates to the successful application of vitrification technology. In recent decades considerable attention has focused on the effects of warming rates in the recovery of embryos cryopreserved by vitrification. In a study by Seki and Mazur (2009), an association was observed between the warming rates of embryos after cryopreservation and embryo viability. Regardless of prior cooling rates the viability of mouse oocytes was linked to a rapid warming rate during their recovery from the frozen state. Mazur and Seki (2011) reported that rapid warming rates are necessary to avoid devitrification during the warming process. Applying the principle of the use of a rapid warming rate, as outlined above Seki, Jin and Mazur (2014) found that under such recovery conditions CPAs can be effective at half the levels used historically. In terms of both exposure time and concentrations used this has the effect of reducing contact of embryonic cells with these potentially cytotoxic cryoprotective agents.

Development of Cryo-Vessels for Vitrification

The development of vessels for cryopreservation of cattle oocytes and embryos offers many options for the application of vitrification methods. The combination of different cryoprotective additives and the use of distinct container types explains the variety of vitrification processes. Nedambale et al. (2004) utilized the VS3a vitrification method, which simply consists of using 6.5 M glycerol with 6% w/v BSA, reporting attainment of 30% pregnancy rates after transferring vitrified embryos into recipient cows. Somfai et al. (2006) defined solid surface vitrification (SSV) as a vitrification method in which there is utilization of a metal surface, pre-cooled to -180 °C by partial immersion in liquid nitrogen, to cool the microdrops of vitrification solution containing ova or embryos. In addition, the SSV solution contains 35% EG, 5% polyvinyl pyrrolidone (PVP) and 0.3 M trehalose. It was reported that the relatively greater concentration of CPAs in SSV solutions may not impair pig oocyte development *in vitro*. Similar to SSV, the use of cryoloop vitrification (CLV) procedures resulted in acceptable survival rates of goat oocytes and embryos (Begin et al., 2003). In this experiment, the SSV solution was modified as follows: 35% EG, 5% PVP and 0.4% trehalose; while the CLV solution contained 20% DMSO, 20% EG, 10 mg/ml Ficoll and 0.65 M sucrose. The CLV vitrification protocol, in which oocytes or embryos are placed in a loop with the aid of a stereomicroscope, seems to be a very intricate procedure to perform. Hochi et al. (2004) indicated that the use of CLV vitrification procedures was not suitable for rabbit zygotes because after frozen/thawing by CLV cleavage rates were only 5% and no embryos developed to blastocyst stages. Furthermore, it was also shown that use of cryotop vitrification of embryos that were placed in a narrow, thin film strip attached to a hard plastic holder resulted in acceptable survival rates of embryos.

Inconsistent with the findings for CLV, however, is the study of Yacoub, Gauly and Holtz (2010), in which use of an OPS vitrification procedure resulted in superior viability of embryos. It is likely that OPS is a method that will become widely adopted in the field as it permits dilution of the solutions in which the embryos are stored within the glass straw followed by direct transfer into recipients. This supposition is supported by the fact that the OPS procedure may be used to cryopreserve oocytes or embryos of various species. These include cattle oocytes (Vieira et al., 2002), cattle embryos (Mucci et al., 2006), horse embryos (Moussa et al., 2005) and rabbit embryos (Lin et al., 2011). However, Yu et al. (2010) reported that the use of the OPS protocol may lead to potential contamination of cattle embryos due to the direct contact of embryos with liquid nitrogen. In order to avoid this problem, a closed pull straw (CPS) method was devised, a modification of the OPS procedure in which the top end of a glass straw was closed using tweezers (heated in a flame) to prevent embryos coming into contact with liquid nitrogen. Yu et al. (2010) also demonstrated similar embryo survival rates when OPS and CPS were used for vitrification. The cryotop procedure pioneered by Kuwayama (2005), originally developed for human oocytes, has to date not been used widely in the commercial cattle breeding sector. Recently, however, the use of cryotop combined with modifications to the vitrification solution resulted in greater survival rates of *in vitro*-derived cattle embryos when using both *in vitro* and *in vivo* methods for embryo transfer (Do et al., 2017; Do et al., 2018). These promising findings may warrant a reappraisal of the value of this cryopreservation technique for use with cattle embryos.

Enhancement in Characteristics of Embryos Prior to Cryopreservation

In parallel with the improvement of vitrification technologies the optimization of *in vitro* culture media for improvement of embryo viability has enhanced outcomes when using cryopreserved embryos for transfer into recipient animals (Do et al., 2019). It is widely recognized that cattle embryos which are derived using *in vitro* procedures are more sensitive to cooling than are those recovered *in vivo* (Mahmoudzadeh et al., 1994; Al Darwich et al., 2010). The reason for this is thought to be an abnormal accumulation of lipids in the cytoplasm of *in vitro*-derived embryos (Pereira et al., 2008).

Seidel (2006) recommended that in order to improve the cryotolerance of oocytes and embryos it is necessary to enhance their quality prior to cryopreservation by modifications to *in vitro* culture systems. Nedambale et al. (2004) reported that the characteristics of the culture media for *in vitro*-derived cattle embryos could have pronounced effects on blastocyst development rate and embryo viability following cryopreservation. Additionally, Pereira et al. (2008) reported that supplementing *in vitro* culture media with trans-10, cis-12 conjugated linoleic acid to reduce embryonic cytoplasmic lipid deposition results in increased viability of embryos after cryopreservation and thawing for subsequent transfer into recipient cows. Furthermore, Nedambale et al. (2004) demonstrated that co-culturing cattle embryos with synthetic oviductal fluid (SOF) and fetal calf serum (FCS) achieved 30% pregnancy rates after transferring vitrified embryos to recipients. Gómez et al. (2008), however, concluded that there was a greater hatching rate after vitrification/warming when using expanded embryos derived from *in vitro* culture supplemented with BSA (regarded as serum-free culture medium) than for similar embryos cultured in SOF + 5% FCS medium.

CONCLUSION

Globally, in recent times considerable and concerted efforts have been made to conserve cattle genetic resources. *In situ* and *ex situ* conservation programs have been implemented on an international scale, although among countries there are vast differences in the strategies adopted to protect indigenous cattle from extinction. Less developed countries focus much more on the inclusion of animals from breeds at risk of local extinction in cattle herds maintained for breed conservation purposes. In contrast, more developed nations invest in both *in situ* conservation and the establishment of gene banks for the storage of cattle genetic materials. Establishing gene banks is a very important step towards securing cattle genetic resources in the long-term, and this depends on the capacity of the use of cryopreservation protocols. Cryopreservation techniques have evolved rapidly since the turn of this century. Application of the traditional slow freezing method remains important to preservation of cattle semen and embryos. Meanwhile, there has been a major improvement in ease of use and successful outcomes of vitrification procedures since the procedure was first developed more than three decades ago. It is envisaged that this technique will be used more routinely for the purposes of cattle breed conservation in years to come.

CONFLICTS OF INTEREST

The authors declare no conflicts of interest.

Acknowledgments

The authors' research reported herein was supported financially by Central Queensland University and Australian Reproductive Technologies. Van Huong Do is in receipt of an Australia Awards Scholarship.

References

Al Darwich, A. A., Perreau, C., Petit, M. H., Papillier, P., Dupont, J., Guillaume, D., Mermillod, P. and Guignot, F. (2010). Effect of PUFA on embryo cryoresistance, gene expression and AMPKα phosphorylation in IVF-derived bovine embryos. *Prostaglandins and Other Lipid Mediators* 93, 30-36.

Asada, M., Ishibashi, S. Ikumi, S. and Fukui, Y. (2002). Effect of polyvinyl alcohol (PVA) concentration during vitrification of *in vitro* matured bovine oocytes. *Theriogenology* 58, 1199-1208.

Begin, I., Bhatia, B., Baldassarre, H., Dinnyes, A. and Keefer, C. L. (2003). Cryopreservation of goat oocytes and *in vivo* derived 2- to 4-cell embryos using the cryoloop (CLV) and solid-surface vitrification (SSV) methods. *Theriogenology* 59, 1839-1850.

Berlinguer, F., Succu, S., Mossa, F., Madeddu, M., Bebbere, D., Leoni, G. G. and Naitana, S. (2007). Effects of trehalose co-incubation on *in vitro* matured prepubertal ovine oocyte vitrification. *Cryobiology* 55, 27-34.

Blackburn, H. D. (2009). Genebank development for the conservation of livestock genetic resources in the United State of America. *Livestock Science* 120, 196-203.

Campos-Chillòn, L. F., Walker, D. J., de la Torre-Sanchez, J. F. and Seidel, J. G. E. (2006). *In vitro* assessment of a direct transfer vitrification procedure for bovine embryos. *Theriogenology* 65, 1200-1214.

Checura, C. M. and Seidel, J. G. E. (2007). Effect of macromolecules in solutions for vitrification of mature bovine oocytes. *Theriogenology* 67, 919-930.

Cuello, C., Gil, M. A., Almiñana, C., Sanchez-Osorio, J., Parrilla, I., Caballero, I., Vazquez, J. M., Roca, J., Rodriguez-Martinez, H. and Martinez, E. A. (2007). Vitrification of *in vitro* cultured porcine two-to-four cell embryos. *Theriogenology* 68, 258-264.

Danchin-Burge, C., Hiemstra, S. J. and Blackburn, H. (2011). *Ex situ* conservation of Holstein-Friesian cattle: comparing the Dutch, French, and US germplasm collections. *Journal of Dairy Science* 94, 4100-4108.

Do, V. H., Walton, S., Catt, S. and Taylor-Robinson, A. W. (2017). A comparative analysis of the efficacy of three cryopreservation protocols on the survival of *in vitro*-derived cattle embryos at pronuclear and blastocyst stages. *Cryobiology* 77, 58-63.

Do, V. H. Catt, S., Amaya, G., Batsiokis, M., Walton, S. and Taylor-Robinson, A. W. (2018). Comparison of pregnancy when non-vitrified and vitrified *in vitro*-derived embryos are transferred into recipients. *Theriogenology* 120, 105-110.

Do, V. H., Catt, S., Kinder, J. E., Walton S. and Taylor-Robinson A. W. (2019). Vitrification of *in vitro*-derived cattle embryos: targeting enhancement of quality by refining technology and standardizing procedures. *Reproduction, Fertility and Development* 31, 837-846.

Dobrinsky, J. R. (2001). Cryopreservation of swine embryos: a chilly past with a vitrifying future. *Theriogenology* 56, 1333-1344.

Fan, W. X., Ma, X. H., Ge, D., Liu, T. Q. and Cui, Z. F. (2009). Cryoprotectants for the vitrification of corneal endothelial cells. *Cryobiology* 58, 28-36.

FAO: Food and Agriculture Organization of the United Nations (2013). *In Vivo Conservation of Animal Genetic Resources* (No. 14). FAO Animal Production and Health Guidelines. http://www.fao.org/3/i3327e/i3327e.pdf.

FAO: Food and Agriculture Organization of the United Nations (2014). Commission on Genetic Resources for Food and Agriculture. Status and Trends of Animal Genetic Resources. http://www.fao.org/3/a-at135e.pdf.

FAO: Food and Agriculture Organization of the United Nations (2015). The Second Report on the State of the World's Animal Genetic Resources for Food and Agriculture. FAO Commission on Genetic Resources for Food and Agriculture Assessments. http://www.fao.org/3/a-i4787e.pdf.

Gómez, E., Rodríguez, A., Muñoz, M., Caamaño, J. N., Hidalgo, C. O., Morán, E., Facal, N. and Díez, C. (2008). Serum free embryo culture medium improves *in vitro* survival of bovine blastocysts to vitrification. *Theriogenology* 69, 1013-1021.

Hochi, S., Terao, T., Kamei, M., Kato, M., Hirabayashi, M. and Hirao, M. (2004). Successful vitrification of pronuclear-stage rabbit zygotes by minimum volume cooling procedure. *Theriogenology* 61, 267-275.

Hochi, S., Abdalla, H., Hara, H., Shimoda, M., Morita, H., Kuwayama, M. and Hirabayashi, M. (2010). Stimulatory effect of Rho-associated coiled-coil kinase (ROCK) inhibitor on revivability of *in vitro*-produced bovine blastocysts after vitrification. *Theriogenology* 73, 1139-1145.

Ishimori, H., Saeki, K., Inai, M., Nagao, Y., Itasaka, J., Miki, Y., Seike, N. and Kainuma, H. (1993). Vitrification of bovine embryos in a mixture of ethylene glycol and dimethyl sulfoxide. *Theriogenology* 40, 427-433.

Kasprzak-Filipek, K., Sawicka-Zugaj, W., Litwińczuk, Z., Chabuz, W., Šveistiene, R. and Bulla, J. (2018). Assessment of the genetic structure of Central European cattle breeds based on functional gene polymorphism. *Global Ecology and Conservation* 17, e00525.

Kuwayama, M. (2005). Highly efficient vitrification method for cryopreservation of human oocytes. *Reproductive BioMedicine Online* 11, 300-308.

Lauvie, A., Danchin-Burge, C., Audiot, A., Brives, H., Casabianca, F. and Verrier, E. (2008). A controversy about crossbreeding in a conservation programme: the case study of the Flemish Red cattle breed. *Livestock Science* 118, 113-122.

Leibo, S. P. (1989). Equilibrium and nonequilibrium cryopreservation of embryos. *Theriogenology* 31, 85-93.

Leibo, S. P. (2008). Cryopreservation of oocytes and embryos: optimization by theoretical versus empirical analysis. *Theriogenology* 69, 37-47.

Leroy, G., Baumung, R., Boettcher, P., Besbes, B., From T. and Hoffmann, I. (2018). Animal genetic resources diversity and ecosystem services. *Global Food Security* 17, 84-91.

Lin, T. A., Chen, C. H., Sung, L. Y., Carter, M. G., Chen, Y. E., Du, F., Ju, J. C. and Xu, J. (2011). Open-pulled straw vitrification differentiates cryotolerance of *in vitro* cultured rabbit embryos at the eight-cell stage. *Theriogenology* 75, 760-768.

Loutradi, K. E., Kolibianakis, E. M., Venetis, C. A., Papanikolaou, E. G., Pados, G., Bontis, I. and Tarlatzis, B. C. (2008). Cryopreservation of human embryos by vitrification or slow freezing: a systematic review and meta-analysis. *Fertility and Sterility* 90, 186-193.

Mahmoudzadeh, A. R., Van Soom, A., Ysebaert, M. T. and de Kruif, A. (1994). Comparison of two-step vitrification versus controlled freezing on survival of *in vitro* produced cattle embryos. *Theriogenology* 42, 1389-1397.

Martin-Collado, D., Diaz, C., Drunker, A. G. and Carabano, M. J. (2014). Determination of non-market values to inform conservation strategies for the threatened Alistana-Sanabresa cattle breed. *Animal* 8, 1373-1381.

Matsunari, H., Maehara, M., Nakano, K., Ikezawa, Y., Hagiwara, Y., Sasayama, N., Shirasu, A., Ohta, H., Takahashi, M. and Nagashima, H. (2012). Hollow fiber vitrification: a novel method for vitrifying multiple embryos in a single device. *Journal of Reproduction and Development* 58, 175-179.

Mazur, P. and Seki, S. (2011). Survival of mouse oocytes after being cooled in a vitrification solution to -196°C at 95° to 70,000 °C/min and warmed at 610° to 118,000°C/min: a new paradigm for cryopreservation by vitrification. *Cryobiology* 62, 1-7.

Morató, R., Izquierdo, D., Paramio, M. T. and Mogas, T. (2008). Cryotops versus open-pulled straws (OPS) as carriers for the cryopreservation of bovine oocytes: effects on spindle and chromosome configuration and embryo development. *Cryobiology* 57, 137-141.

Moussa, M., Bersinger, I., Doligez, P., Guignot, F., Duchamp, G., Vidament, M., Mermillod, P. and Bruyas, J. F. (2005). *In vitro* comparisons of two cryopreservation techniques for equine embryos: slow-cooling and open pulled straw (OPS) vitrification. *Theriogenology* 64, 1619-1632.

Mucci, N., Aller, J., Kaiser, G. G., Hozbor, F., Cabodevila, J. and Alberio, R. H. (2006). Effect of estrous cow serum during bovine embryo culture on blastocyst development and cryotolerance after slow freezing or vitrification. *Theriogenology* 65, 1551-1562.

Nedambale, T. L., Dinnyés, A., Groen, W., Dobrinsky, J. R., Tian, X. C. and Yang, X. (2004). Comparison of *in vitro* fertilized bovine embryos cultured in KSOM or SOF and cryopreserved by slow freezing or vitrification. *Theriogenology* 62, 437-449.

Nowak, A., Kochan, J., Papis, K. and Okolski, A. (2014). Studies on survival of horse ooctyes after rapid-i method vitrification. *Journal of Equine Veterinary Science* 34, 675-679.

Ochert, A. (2016). *National gene bank opens in China*. https://www.bionews.org.uk/page_95701

Paiva, S. R., McManus, C. M. and Blackburn, H. (2016). Conservation of animal genetic resources – a new tact. *Livestock Science* 193, 32-38.

Palasz, A. T., Gustafsson, H., Rodriguez-Martinez, H., Gusta, L., Larsson, B. and Mapletoft, R. J. (1997). Vitrification of bovine IVF blastocysts in an ethylene glycol/sucrose solution and heat-stable plant-extracted proteins. *Theriogenology* 47, 865-879.

Pereira, R. M., Carvalhais, I., Pimenta, J., Baptista, M. C., Vasques, M. I., Horta, A. E. M., Santos, I. C., Marques, M. R., Reis, A., Pereira, M. S. and Marques, C. C. (2008). Biopsied and vitrified bovine embryos viability is improved by trans10, cis12 conjugated linoleic acid supplementation during *in vitro* embryo culture. *Animal Reproduction Science* 106, 322-332.

Pugh, P. A., Tervit, H. R. and Niemann, H. (2000). Effects of vitrification medium composition on the survival of bovine *in vitro* produced embryos, following in straw-dilution, *in vitro* and *in vivo* following transfer. *Animal Reproduction Science* 58, 9-22.

Rios, G. L., Mucci, N. C., Kaiser, G. G. and Alberio, R. H. (2010). Effect of container, vitrification volume and warming solution on cryosurvival of *in vitro*-produced bovine embryos. *Animal Reproduction Science* 118, 19-24.

Sanches, B. V., Marinho, L. S. R., Filho, B. D. O., Pontes, J. H. F., Basso, A. C., Meirinhos, M. L. G., Silva-Santos, K. C., Ferreira, C. R. and Seneda, M. M. (2013). Cryosurvival and pregnancy rates after exposure of IVF-derived *Bos indicus* embryos to forskolin before vitrification. *Theriogenology* 80, 372-377.

Seidel, G. E. Jr. (2006). Modifying oocytes and embryos to improve their cryopreservation. *Theriogenology* 65, 228-235.

Seki, S. and Mazur, P. (2009). The dominance of warming rate over cooling rate in the survival of mouse oocytes subjected to a vitrification procedure. *Cryobiology* 59, 75-82.

Seki, S., Jin, B. and Mazur, P. (2014). Extreme rapid warming yields high functional survivals of vitrified 8-cell mouse embryos even when suspended in a half-strength vitrification solution and cooled at moderate rates to -196 °C. *Cryobiology* 68, 71-78.

Shirazi, A., Soleimani, M., Karimi, M., Nazari, H., Ahmadi, E. and Heidari, B. (2010). Vitrification of *in vitro* produced ovine embryos at various developmental stages using two methods. *Cryobiology* 60, 204-210.

Somfai, T., Dinnyés, A., Sage, D., Marosán, M., Carnwath, J. W., Ozawa, M., Kikuchi, K. and Niemann, H. (2006). Development to the blastocyst stage of parthenogenetically activated *in vitro* matured porcine oocytes after solid surface vitrification (SSV). *Theriogenology* 66, 415-422.

Taberlet, P., Coissac, E., Pansu, J. and Pompanon, F. (2011). Conservation genetics of cattle, sheep, and goats. *Comptes Rendus Biologies* 334, 247-254.

Tamou, C., de Boer, I. J. M., Ripoll-Bosch, R. and Oosting, S. J. (2018). Understanding roles and functions of cattle breeds for pastoralists in Benin. *Livestock Science* 210, 129-136.

Taylor-Robinson, A. W., Walton, S., Swain, D. L., Walsh, K. B. and Vajta, G. (2014). The potential for modification in cloning and vitrification technology to enhance genetic progress in beef cattle in Northern Australia. *Animal Reproduction Science* 148, 91-96.

Tienhaara, A., Ahtiainen, H. and Pouta, E. (2015). Consumer and citizen roles and motives in the valuation of agricultural genetic resources in Finland. *Ecological Economics* 114, 1-10.

Vajta, G. (2000). Vitrification of the oocytes and embryos of domestic animals. *Animal Reproduction Science* 60-61, 357-364.

Vajta, G., Holm, P., Greve, T. and Callesen, H. (1996). Factors affecting survival rates of *in vitro* produced bovine embryos after vitrification and direct in-straw rehydration. *Animal Reproduction Science* 45, 191-200.

van Wagtendonk-de Leeuw, A. M., den Daas, J. H.G. and Rall, W. F. (1997). Field trial to compare pregnancy rates of bovine embryo cryopreservation methods: vitrification and one-step dilution versus slow freezing and three-step dilution. *Theriogenology* 48, 1071-1084.

Vieira, A. D., Mezzalira, A., Barbieri, D. P., Lehmkuhl, R. C., Rubin, M. I. B. and Vajta, G. (2002). Calves born after open pulled straw vitrification of immature bovine oocytes. *Cryobiology* 45, 91-94.

Yacoub, A. N. A, Gauly, M. and Holtz, W. (2010). Open pulled straw vitrification of goat embryos at various stages of development. *Theriogenology* 73, 1018-1023.

Yu, X. L., Deng, W., Liu, F. J., Li, Y. H., Li, X. X., Zhang, Y. L. and Zan, L. S. (2010). Closed pulled straw vitrification of *in vitro*-produced and *in vivo*-produced bovine embryos. *Theriogenology* 73, 474-479.

Zander, K. K. and Drucker, A. G. (2008). Conserving what's important: using choice model scenarios to value local cattle breeds in East Africa. *Ecological Economics* 68, 34-45.

Zander, K. K., Drucker, A. G. and Holm-Müller, K. (2009). Costing the conservation of animal genetic resources: the case of Borana cattle in Ethiopia and Kenya. *Journal of Arid Environments* 73, 550-556.

Zander, K. K., Signorello, G., Salvo, M. D., Gandini, G. and Drucker, A. G. (2013). Assessing the total economic value of threatened livestock breeds in Italy: implications for conservation policy. *Ecological Economics* 93, 219-229.

In: Agricultural Research Updates. Vol. 28 ISBN: 978-1-53616-253-0
Editors: P. Gorawala, S. Mandhatri

Chapter 8

FUNCTIONS OF *NADPH OXIDASE COMPLEX IN DEVELOPMENT AND PATHOGENICITY OF BOTRYTIS CINEREA*

Shiping Tian*, Zhanquan Zhang, Guozheng Qin, Yong Chen, Boqiang Li and Tong Chen
Key Laboratory of Plant Resources, Institute of Botany, The Innovative Academy of Seed Design, Chinese Academy of Sciences, Beijing, China

ABSTRACT

Reactive oxygen species (ROS) play crucial roles in the defense and differentiation of multi-cellular organisms. Nicotinamide adenine dinucleotide phosphate (NADPH) oxidase complex serves as the major enzymatic producer of ROS in eukaryotic cell. The NADPH oxidase (Nox) complex has been found in many fungi, and its functions have been well studied in some plant pathogenic fungi. *Botrytis cinerea* is one of the most devastating plant pathogenic fungi, and represents a model organism to study the pathogenicity of aggressive fungal pathogens due to its broad

* Corresponding Author's E-mail: tsp@ibcas.ac.cn.

host range and great economic importance. Much work has shown that Nox complex is involved in many biological processes of *B. cinerea*. Here, we will summarize the following research progresses: (i) the structure and components of Nox complex; (ii) functions of the components of Nox complex; (iii) control strategy for *B. cinerea*.

1. INTRODUCTION

Botrytis cinerea, a necrotrophic phytopathogen, causes grey mould disease in more than 500 economically important hosts and is ranked as the second most devastating plant pathogen (Dean et al., 2012) This pathogen serves as the most important postharvest pathogen and is destructive on mature or senescent fruits, including pome fruit, stone fruit and berry fruit, to cause astonishing economic losses annually (Williamson et al., 2007). Moreover, *B. cinerea* is difficult to be controlled in practice, because it has strong resistance to adversity, a variety of modes of attack, diverse hosts as inoculum sources, strong drug resistant and genetic plasticity (Fillinger and Elad, 2016). So far, application of any single control measure is unlikely to succeed. Due to its economic importance and proven technique of genetic transformation, *B. cinerea* has become one of the useful model systems for studying the pathogenesis and other biological issues of aggressive fungal pathogens. During the past decades, great efforts have been made to unravel the molecular mechanisms of the pathogenicity of *B. cinerea* and many key genes regulating pathogenicity have been identified in our laboratory (Figure 1). The pathogenicity of *B. cinerea* is regulated by different factors, including environmental conditions, signal transduction, protein secretion, and reactive oxygen species (ROS) and so on. During infection process, *B. cinerea* can secrete an arsenal of enzymes to facilitate its colonization, small GTPase Bcsas1 has been proved to affect the development and pathogenicity of *B. cinerea* via regulating the protein secretion (Zhang et al., 2014). The ambient pH can adjust the composition of the secretome to affect the virulence of *B. cinerea*. At pH4, the secretome of *B. cinerea* was rich in proteolysis, whereas more cell wall–degrading enzymes (CWDEs) were detected in the secretome at higher pH value (pH 6) to achieve full virulence (Li et al., 2012). The light responsive transcriptional factor, Bcmads1, was involved in the photomorphogenesis and pathogenicity of *B. cinerea* (Zhang et al., 2016). Recently, ROS generating systems and some components have been proved to be involved in the development and pathogenicity of *B. cinerea* (An et al., 2015; Li et al., 2016; Li et al., 2018).

ROS are highly reactive molecules, including superoxide anion ($\bullet O_2^-$), hydroxyl radical ($\bullet OH$) and hydrogen (H_2O_2), and ubiquitous in living cells. ROS, as common and unavoidable byproducts of some metabolic processes in organisms, play a dual role in the interaction between microbe and their hosts (Heller and Tudzynski, 2011; Tian et al., 2013). When attacked by pathogen, the plant cells were able to produce amount of ROS transiently, the so-called "oxidative burst", as the resistance response (Wojtaszek 1997; Torres et al., 2002). On the one hand, ROS produced by plant hosts could lead to the crosslinking of cell wall structural proteins and the hypersensitive response to limit the spread of infection (Brisson et al., 1994; Levine et al., 1994); On the other hand, ROS could cause the damage of biological macromolecules, such as protein oxidation and DNA mutation, thus inhibiting the activity of pathogen (Reverter-Branchat et al., 2004; Branduardi et al., 2007). Meanwhile, fungal pathogens are also able to produce ROS to achieve differentiation and full pathogenicity. ROS are usually generated in mitochondria, cytochrome P-450, peroxisomes and other cellular components (Heller and Tudzynski, 2011). The mitochondria are considered to produce continuous levels of $\bullet O_2^-$ at complex I and III of electron transport chain (Qin et al., 2011). Nox complex is the most common enzymatic system to produce ROS, and generates superoxide anion ($\bullet O_2^-$) through a one-electron reduction of molecular oxygen, using electrons supplied by NADPH (Bedard and Krause 2007; Takemoto et al. 2011). Then $\bullet O_2^-$ is converted to H_2O_2 via superoxide dismutase (SOD), and H_2O_2 is further converted to other forms of ROS. Nox has been found in many fungi and its functions have been well studied in *B. cinerea*.

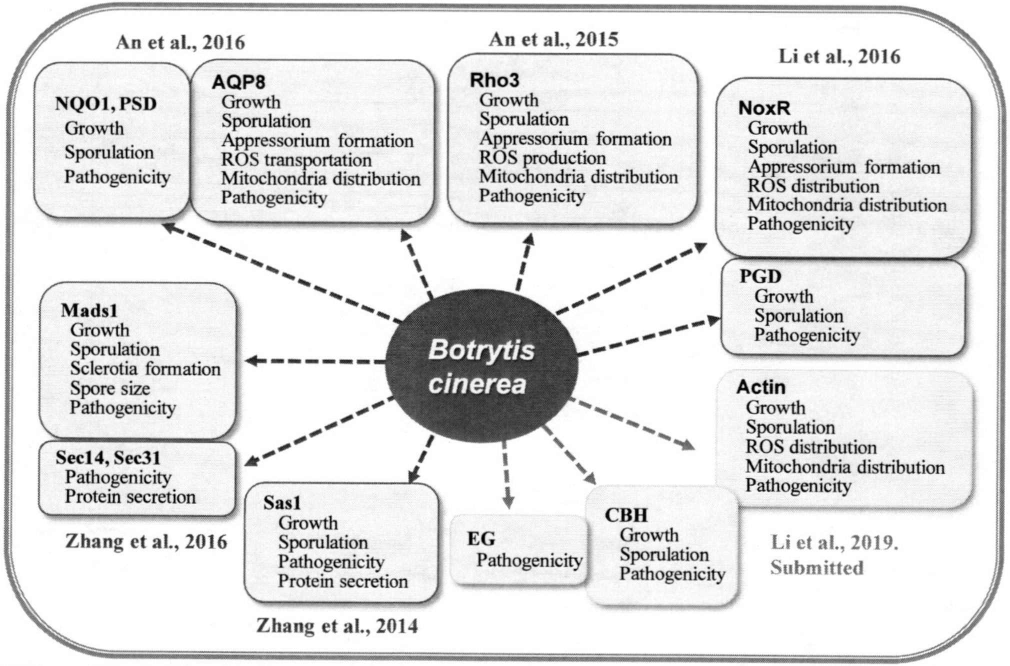

Sas1: Rab family GTPase, **Rho3**: Rho family GTPase, **Mads1**: MADS-box transcription factor, **Sec14**: cytosolic factor that catalyzes the transfer of phosphatidylinositol and phosphatidylcholine, **Sec31**: component of COPII coat of vesicles, **AQP8**: aquaporin, **NQO1**: quinone oxidoreductase, **PSD**: phosphatidylserine decarboxylase, **NoxR**: regulatory subunit of NADPH oxidase, **PGD**: 6-phosphogluconate dehydrogenase, **Actin**: component of microfilaments, **EG**: endoglucanase, **CBH**: β-d-glucan cellobiohydrolase

Figure 1. The pathogenic genes identified in *B. cinerea*.

In this chapter, we will summarize the research progress of Nox complex in *B. cinerea*, including the structure of Nox complex and the functions of Nox complex in regulating ROS production, differentiation and pathogenicity of *B. cinerea*, as well as control strategy for *B. cinerea*.

2. The Structure and Components of Nox Complex in *B. cinerea*

The Nox complexes have been comprehensively characterized in mammalian and human cells. Nox2 in humans, comprising six different subunits that interact with each other to form an active enzyme complex, was responsible for the production of superoxide (Babior, 1999; Panday et al., 2015). Two Nox subunits, gp91^{phox} and p22^{phox}, are integral membrane proteins that together form the large heterodimeric subunit flavocytochrome b558 (Figure 2A). In the resting stage, the multi-domain regulatory subunits, p40^{phox}, p47^{phox} and p67^{phox}, existed in the cytosol as a complex (Groemping and Rittinger, 2005). Upon cell stimulation, the cytosolic complex was recruited from the cytosol to the membrane where they interacted with gp91^{phox} and p22^{phox} through a phosphorylation-induced conformational change of p47^{phox} (Han et al., 1998; Groemping et al., 2003). The activation of the complex also required the small GTPase Rac1/2, which acted as a molecular switch for Nox activation through binding in the GTP form to p67^{phox} (Diebold et al., 2001). Once assembled, the complex was active and generates superoxide anion by transferring electrons from NADPH in the cytosol to dioxygen on the luminal or extracellular space (Koga et al., 1999).

There are quite a few homologies between fungal and mammalian Nox complex. As most other fungi, *B. cinerea* possesses two catalytic subunits, BcNoxA (Figure 2B) and BcNoxB (Figure 2C), both of them are homologs of the human catalytic subunits gp91^{phox} and possess the core Nox transmembrane and catalytic domains (Takemoto et al. 2007; Segmüller et al., 2008). Recently, a homolog of the mammalian adaptor protein p22^{phox}, BcNoxD, was identified to directly interact with the catalytic subunit BcNoxA (Siegmund et al., 2013). Another transmembrane protein, BcPls1, took over the p22^{phox} function in BcNoxB complexes (Siegmund et al., 2013). In *B. cinerea*, the catalytic function of both BcNoxA and BcNoxB were regulated by BcNoxR, which may be a homologue of the mammalian regulatory subunit p67^{phox} (Segmüller et al., 2008; Siegmund et al., 2013). Nevertheless, the direct interaction between BcNoxR and BcNoxA/B has so far not been confirmed. A scaffold protein, BcIqg1, has been recently identified to mediate the interaction of the catalytic subunits with the regulator BcNoxR by interacting with BcNoxD and the dominant active form of BcRac which was shown to interact with BcNoxR (Marschall and Tudzynski, 2016). However, the homologues of p40^{phox} and p47^{phox} are absence in fungal genome databases. In *Epichloë festucae*, two well-characterized regulators of yeast polarity proteins, BemA and Cdc24, were identified to be components of the fungal Nox-complex since they interacted with each another and with NoxR (Takemoto et al., 2011).

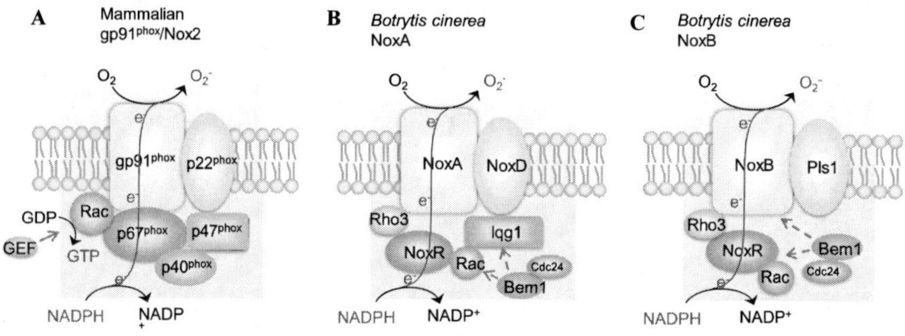

Figure 2. Proposed composition of the *B. cinerea* NoxA/B complex and comparison with the known composition of the mammalian Nox2 complex. The schematic is modified from Scott (2015).

Comparable studies in *B. cinerea* did not show a similar interaction of Bem1 and Cdc24 with NoxR. Only BcCdc24 and BcBem1 strongly interacted with each other and the Δ*BcBem1* showed the similar phenotype as Δ*BcNoxA/B* (Giesbert et al., 2014). In addition, a small GTPase BcRho3 was also proved to be a component of the fungal Nox-complex since it interacted with BcNoxA/B and NoxR, and it was related to mitochondria distribution and intracellular ROS generation (An et al., 2015). Nox in *B. cinerea* might function at two different locations. The BcNoxA complex can transfer electrons through the ER and has a predominant role inside the ER, while the BcNoxB complex was mainly important outside the ER, likely at the plasma membrane (Siegmund et al., 2013; Marschall et al., 2016).

3. Functions of the Components of Nox Complex in *B. cinerea*

3.1. Nox Complex Regulates ROS Production and Distribution

Nox complex is the major enzymatic systems to produce ROS in eukaryotic cell. In *B. cinerea*, a lot of efforts were made to study the functions of Nox complex during the last decade. Although the subunits of Nox complex are involved in various biological processes in *B. cinerea*, the ROS accumulation in cell of *B. cinerea* is not greatly influenced by the subunits of Nox complex (Fillinger and Elad, 2016). The Nox mutants, including *ΔbcnoxA*, *ΔbcnoxB*, *ΔbcnoxAB*, *ΔbcnoxR* and *Δbcpls*, did not show reduced ROS level compared to wild-type strain and the *ΔbcnoxA* mutant even showed an increased production of H_2O_2 (Segmüller et al., 2008; Siegmund et al., 2013). In further, the ROS levels of the Nox deletion mutants during infection were detected by nitro blue tetrazolium (NBT) staining. The results indicated that there was no any detectable difference in extracellular ROS levels between the mutants and wild type (Segmüller et al., 2008). These results suggested that although Nox complex played an important role in regulating the differentiation and pathogenicity of *B. cinerea*, it did not mainly contribute to the ROS production. Recent study of Li et al. (2016) found that Nox complexes had important influence on the polar distribution of ROS at the hyphae tips of *B. cinerea*. They compared the distribution of ROS and mitochondria at the hyphae tips between wild type and *BcnoxR* mutant by staining with DCFH-DA (ROS indicator) and Mito-Tracker orange (a fluorescent dye specific for mitochondria), respectively, and found that ROS polar distributed in wild type, while exhibited even distribution in *BcnoxR* mutant (Li et al., 2016). Mitochondrial location concurred with ROS distribution both in WT and mutant. These results demonstrated that BcnoxR could influence the polar distribution of ROS and mitochondrial location in *B. cinerea*. Previous studies showed that small GTPase could regulate the ROS generation by Nox complex in animal, plant, and fungi (Bedard and Krause, 2007; Tanaka et al., 2008; Fluhr, 2009; Takemoto et al., 2011). In *B. cinerea*, a small GTPase Rho3 is a potential subunit of Nox complex and can interact with NoxA, NoxB and NoxR (Unpublished data). The disruption of *rho3* also resulted in the disappearance of polar distribution of intercellular ROS at hyphae tips (An et al., 2015). The distribution pattern of ROS at the hyphae tips is important for polar growth of fungi. The extracellular ROS generated by Nox complex might be transported back to the interior of cells by the water-channel protein aquaporin, which located at plasma membrane (An et al., 2016). Similar to NoxR and Rho3, aquatorin8 was also involved in the regulation of ROS distribution in *B. cinerea* (An et al., 2015; 2016). These results implied that Nox complex was not engaged in major ROS production in *B. cinerea*, but rather contributed to signal transduction and differentiation through influencing the distribution of ROS in cells or other mechanisms.

3.2. Nox Complex Affects Development of *B. cinerea*

Function analyses of the genes in Nox complex have unraveled that Nox subunits are extensively contributed in differentiation and developmental processes of *B. cinerea* (Table 1). Both BcNoxA and BcNoxB were involved in sclerotia formation, but neither was required for ascospore germination in *B. cinerea* (Segmüller et al., 2008). The similar result was discovered in *Neurospora crassa* and *Podospora anserine* (Malagnac et al., 2004; Cano-Domínguez et al., 2008). In addition,

BcNoxA was responsible for the conidial anastomosis tubes (CATs) fusion (Roca et al. 2012). BcNoxR was the regulatory subunit of BcNoxA and BcNoxB, and the mutant of *BcNoxR* mutant showed an additive phenotype of *BcNoxA* and *BcNoxB* (Segmüller et al., 2008; Li et al., 2016). The mutant of *BcNoxR* was impaired in sclerotia and CATs fusion and more susceptible to oxidative stress (Segmüller et al., 2008; Li et al., 2016). Some studies have showed that another regulatory subunit of Nox, small GTPase BcRac, can regulate the polar growth and cell-cycle progression through mitosis (Minz Dub et al., 2013). It was found that knockout mutant of *BcRac* showed lower growth rate with hyper-branching and deformed hyphae, and was vegetatively sterile (Minz Dub et al., 2013). Rho3 is responsible for differentiation processes of *B. cinerea*. Deletion of *rho3* led to significant decrease of vegetative growth and extension of distance between septa (An et al., 2015). In addition, the formation of sclerotia and conidia was suppressed in deficient mutant *Δrho3* (An et al., 2015).

Table 1. Functions of the components of Nox complex in developmental processes of *B. cinerea*

Subunits	Conidia	Sclerotia	CATs	Functional appressoria	Infection cushion	Vegetative growth
BcNoxA		√	√			
BcNoxB		√				
BcNoxR		√	√			
BcNoxD	√	√	√		√	
BcPIs1		√		√		
BcRac	√					√
BcRho3	√	√				√
BcBem1	√		√			

* The symbol "√" indicates that the subunit is involved in the developmental process.

BcNoxD and tetraspanin BcPls1 are two stabilizing proteins of Nox complex. In *B. cinerea* strains T4 and B.05.10, BcPls1 was found to be related to appressoria-mediated penetration (Gourgues et al., 2004), and also was necessary for female fertility in B05.10 (Siegmund et al., 2013). BcNoxD has high homology to the ER protein Pro41 in *Sordaria macrospora*, and showed similar function to the catalytic subunit BcNoxA in differentiation (Siegmund et al., 2015). BcNoxD could interact with BcNoxA to be involved in the fusion of CATs and formation of conidia and sclerotia (Siegmund et al., 2015). The potential scaffold protein BcBem1 has been proved to play an important role in conidia formation and germination, as well as in establishment of CATs (Giesbert et al. 2014).

3.3. Nox Complex Mediates Pathogenicity of *B. cinerea*

Nox complex plays a crucial role in the regulation of pathogenicity of *B. cinerea*. All subunits of Nox complex found in *B. cinerea* had important influence on the pathogenesis, and possible mechanisms were also speculated. Among the subunits of Nox complex, BcRac and BcPls1 had the most significant effect on virulence of *B. cinerea*. The small GTPase BcRac was necessary for pathogenicity of *B. cinerea*. Deleting *BcRac* resulted in vegetatively sterile and nonpathogenic (Minz Dub et al., 2013). Similarly, the knockout mutant of *BcPls1* was not pathogenic on intact plant tissues, although it could differentiate appressoria as wild type. Expression analysis suggested that *BcPls1* was specifically expressed in conidia, germ tubes and appressoria during host penetration (Gourgues et al., 2004). Both the two catalytic subnuits, BcNoxA and BcNoxB, are essential for the pathogenicity of *B. cinerea*. BcNoxB was responsible for appressoria-mediated penetration, and the deletion mutant of *BcNoxB* showed retarded formation of primary lesions. By contrast, BcNoxA was responsible for the subsequent colonization process of *B. cinerea*. The knockout mutant of *BcNoxA* was able to penetrate the host cell as the wild type, but colonization rate was much lower in the host tissue. The mutant of *BcNoxR* had the identical phenotype of the double mutant of *BcNoxA* and *BcNoxB*, implying that BcNoxR could activate both catalytic subunits (Segmüller et al., 2008). Deletion of *BcNoxR* led to slower growth and sporulation and impaired virulence on apple and tomato fruits as compared to wild type (Li et al., 2016). Based on the comparative proteomic approach, a total of 49 proteins, which are mainly involved in stress response, carbohydrate metabolism, translation and intracellular signaling, were identified as the potential downstream targets of BcNoxR. Among these proteins, 6-phosphogluconate dehydrogenase (6-PGD) and actin have been proved to be closely related to development and pathogenicity of *B. cinerea*. 6-PGD catalyzes the oxidative decarboxylation of 6-phosphogluconate to ribulose 5-phosphate, with a concomitant reduction of NADP to NADPH. The deletion of 6-PGD gene *Bcpgd* reduced the virulence of *B. cinerea* on different hosts, including apple, tomato and strawberry fruits, as well as tomato leaves (Li et al., 2016). Siegmund et al. (2015) reported that BcNoxD could interact with BcNoxA, and showed the similar function as BcNoxA in regulating pathogenicity of *B. cinerea*. Small GTPase Rho3 is also related to the virulence of *B. cinerea*. The *Δrho3* mutant of *B. cinerea* lost the ability to penetrate onion epidermis, and showed significantly reduced virulence to apple and tomato fruits, as well as tomato leaves compared to wild-type ~~strain~~ (An et al., 2015). In other fungi, the scaffold proteins Bem1 and Cdc24 (guanine nucleotide exchange factor, GEF) were the components of Nox complex, and could directly interact with NoxR (Takemoto et al., 2011; Schürmann et al., 2013; Herrmann et al., 2014). In *B. cinerea*, BcBem1 and BcCdc24 have been proved to interact with each other (Giesbert et al. 2014). They reported that the former was involved in the penetration and host colonization, whereas the latter was related to penetration process. Moreover, deletion of *BcCdc24* was lethal and the heterokaryotic mutant showed a similar germination defect as *BcBem1* mutant (Giesbert et al. 2014).

4. Control Strategy for *B. cinerea*

A lot of studies have shown that induction or elimination of intracellular ROS is an important strategy for the control of postharvest diseases. Exogenous application of inducers, such as chitosan, salicylic acid (SA), 1-methylcyclopropene (1-MCP), and methyl jasmonate (MeJA), has been proved to have positive effect on disease control in various fruits, via regulating ROS production and inducing resistance of fruit hosts (Liu et al., 2007, Xu and Tian, 2008; Zhu and Tian, 2012; Zhang et al., 2012; Li et al., 2018). ROS production usually leads to hypersensitive response (HR) in plant hosts, which is a rapid tissue necrosis at the site of infection to delimit the pathogen growth (Durrant and Dong, 2004).

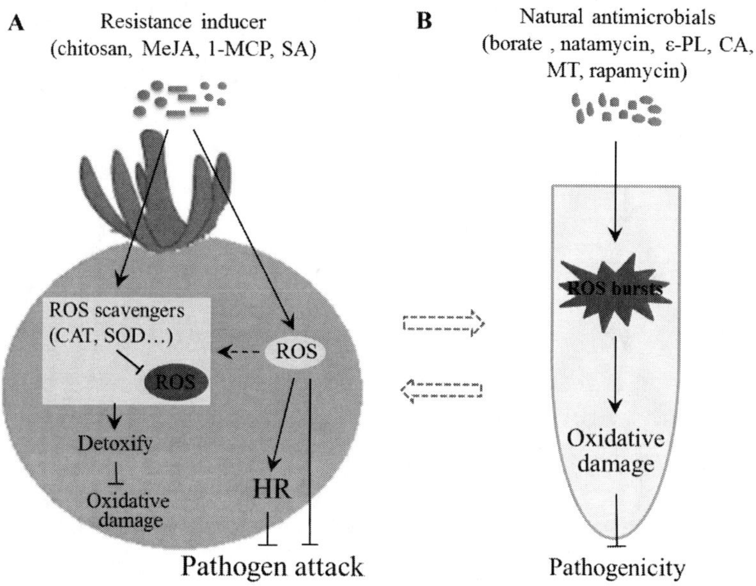

SOD: superoxide dismutase, CAT: catalase, HR: hypersensitive response, MeJA: methyl jasmonate, 1-MCP: 1-methylcyclopropene, SA: salicylic acid, ε-PL: epsilon-polylysine, CA: cinnamic acid, MT: methyl thujate.

Figure 3. The roles of ROS in pathogenesis of fungal pathogens.

On the one hand, the resistance inducers also can stimulate expression of some antioxidant enzymes and relative genes, such as superoxide dismutase (SOD) and catalase (CAT), which are beneficial for scavenging excess ROS and alleviating oxidative damage of fruits (Figure 3A) (Apel and Hirt, 2004; Zhu and Tian, 2012). On the other hand, utilizing of natural antimicrobial to control postharvest disease is also a potential technology instead of chemical fungicides. A variety of biological substances, such as borate, natamycin, cinnamic acid (CA), methyl thujate (MT), rapamycin and epsilon-polylysine (ε-PL) have been reported to effectively inhibit postharvest disease in fruit hosts, including apple, strawberry and tomato fruits by regulating ROS production and accumulation in fungal cell (Figure 3B) (Qin et al., 2007; Zhang et al., 2015; Ji et al., 2018; Li et al., 2019; Ma et al., 2019; He et al., 2019). ROS accumulation in the cell of pathogens can cause oxidative damage and protein carbonylation, resulting in mitochondrial structure destruction (Lai et al., 2011; Qin et al., 2011; Cao et al., 2012). This damage leads to dysfunction of cell function and suppression of fungal growth and pathogenicity (Tian et al., 2013).

CONCLUSION

In brief, various subunits of Nox complex have been identified in *B. cinerea* to date, including two catalytic subnuits BcNoxA and BcNoxB, the regulatory subunits BcNoxR, BcRac and rho3, the stabilizing protein BcPls1 and BcNoxD, and the putative subunits BcBem1 and BcCdc24.

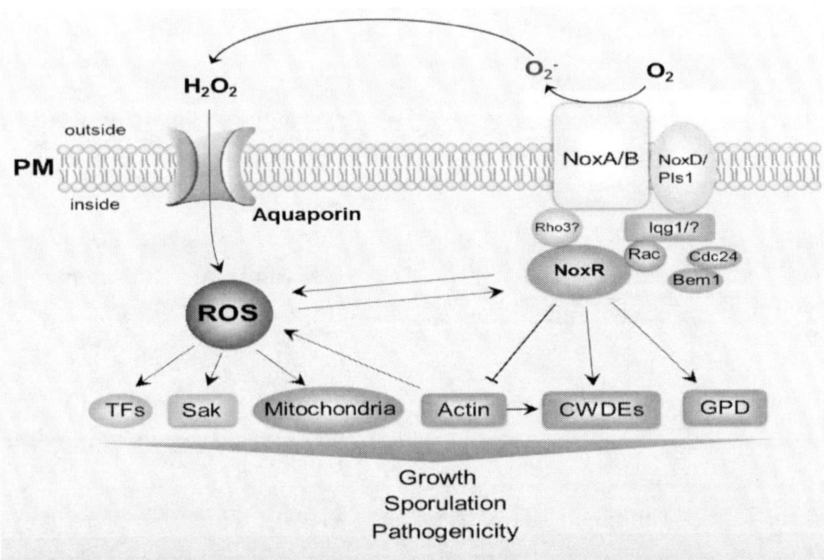

Figure 4. Hypothetical model for NADPH oxidase in the regulation of development and virulence in *B. cinerea*.

The function of Nox complex in regulating the virulence of *B. cinerea* is achieved by mediating the polarity distribution of ROS at the tip of hypha. The results of existing researches have shown that all the subunits of Nox complex have important effect on development and pathogenicity of *B. cinerea*, and each subunit has its own specific function. Synthesis of recent research results, we outline the model for Nox complex in the regulation of development and virulence in *B. cinerea* as Figure 4, showing the structure and components of Nox complex, ROS production and transportation in cell, as well as the action targets of ROS.

REFERENCES

An, B., Li, B. Q., Li, H., Zhang, Z. Q., Qin, G. Z. & Tian, S. P. (2016). Aquaporin8 regulates cellular development and reactive oxygen species production, a critical component of virulence in *Botrytis cinerea*. *New Phytol*, *209*, 1668-1680.

An, B., Li, B. Q., Qin, G. Z. & Tian, S. P. (2015). Function of small GTPase Rho3 in regulating growth, conidiation and virulence of *Botrytis cinerea*. *Fungal Genet Biol*, *75*, 46-55.

Apel, K. & Hirt, H. (2004). Reactive oxygen species: metabolism, oxidative stress, and signal transduction. *Annu Rev Plant Biol*, *55*, 373-399.

Babior, B. M. (1999). NADPH oxidase: an update. *Blood*, *93*, 1464-1476.

Bedard, K. & Krause, K. H. (2007). The NOX family of ROS-generating NADPH oxidases: physiology and pathophysiology. *Physiol Rev*, *87*, 245-313.

Branduardi, P., Fossati, T., Sauer, M., Pagani, R., Mattanovich, D. & Porro, D. (2007). Biosynthesis of vitamin C by yeast leads to increased stress resistance. *PLoS One*, *2*, e1092.

Brisson, L. F., Tenhaken, R. & Lamb, C. (1994). Function of oxidative cross-linking of cell wall structural proteins in plant disease resistance. *Plant Cell*, *6*, 1703-1712.

Cano-Dominguez, N., Alvarez-Delfin, K., Hansberg, W. & Aguirre, J. (2008). NADPH oxidases NOX-1 and NOX-2 require the regulatory subunit NOR-1 to control cell differentiation and growth in *Neurospora crassa*. *Eukaryot Cell*, *7*, 1352-1361.

Cao, B. H., Liu, J., Qin, G. Z. & Tian, S. P. (2012). Oxidative stress acts on special membrane proteins to reduce the viability of *Pseudomonas syringae* pv tomato. *J Proteome Res*, *11*, 4927-4938.

Dean, R., van Kan, J. A., Pretorius, Z. A., Hammond-Kosack, K. E., Pietro, A. D., Spanu, P. D., Rudd, J. J., Dickman, M., Kahmann, R., Ellis, J. & Foster, G. D. (2012). The top 10 fungal pathogens in molecular plant pathology. *Mol Plant Pathol*, *13*, 414-430.

Diebold, B. A. & Bokoch, G. M. (2001). Molecular basis for Rac2 regulation of phagocyte NADPH oxidase. *Nat Immunol*, *2*, 211-215.

Durrant, W. E. & Dong, X. N. (2004). Systemic acquired resistance. *Ann Rev Phytopathol*, *42*, 185-209.

Fillinger, S. & Elad, Y. (2016). *Botrytis*-the fungus, the pathogen and its management in agricultural systems. New York, Springer.

Fluhr, R. (2009). *Reactive oxygen-generating NADPH oxidases in plants*. Berlin, Springer.

Giesbert, S., Siegmund, U., Schumacher, J., Kokkelink, L. & Tudzynski, P. (2014). Functional analysis of BcBem1 and its interaction partners in *Botrytis cinerea*: impact on differentiation and virulence. *PLoS One*, *9*, e95172.

Gourgues, M., Brunet-Simon, A., Lebrun, M. H. & Levis, C. (2004). The tetraspanin BcPls1 is required for appressorium-mediated penetration of *Botrytis cinerea* into host plant leaves. *Mol Microbiol*, *51*, 619-629.

Groemping, Y., Lapouge, K., Smerdon, S. J. & Rittinger, K. (2003). Molecular basis of phosphorylation-induced activation of the NADPH oxidase. *Cell*, *113*, 343-355.

Groemping, Y. & Rittinger, K. (2005). Activation and assembly of the NADPH oxidase: a structural perspective. *Biochem J*, *386*, 401-416.

Han, C. H., Freeman, J. L., Lee, T., Motalebi, S. A. & Lambeth, J. D. (1998). Regulation of the neutrophil respiratory burst oxidase.

Identification of an activation domain in p67^{phox}. *J Biol Chem*, *273*, 16663-16668.

He, C., Zhang, Z. Q., Li, B. Q., Xu, Y. & Tian, S. P. (2019). Effect of natamycin on *Botrytis cinerea* and *Penicillium expansum*-Postharvest pathogens of grape berries and jujube fruit. *Postharvest Biol Technol*, *151*, 134-141.

Heller, J. & Tudzynski, P. (2011). Reactive oxygen species in phytopathogenic fungi: signaling, development, and disease. *Annu Rev Phytopathol*, *49*, 369-390.

Herrmann, A., Tillmann, B., Schürmann, J., Bolker, M. & Tudzynski, P. (2014). Small GTPase-associated signaling by the guanine nucleotide exchange factors CpDock180 and CpCdc24, the GTPase effector CpSte20 and the scaffold CpBem1 in *Claviceps purpurea*. *Eukaryot Cell*, *13*, 470-482.

Ji, D. C., Chen, T., Ma, D. Y., Liu, J. L., Xu, Y. & Tian, S. P. (2018). Inhibitory effects of methyl thujate on mycelial growth of *Botrytis cinerea* and possible mechanisms. *Postharvest Biol Technol*, *142*, 46-54.

Koga, H., Terasawa, H., Nunoi, H., Takeshige, K., Inagaki, F. & Sumimoto, H. (1999). Tetratricopeptide repeat (TPR) motifs of p67^{phox} participate in interaction with the small GTPase Rac and activation of the phagocyte NADPH oxidase. *J Biol Chem*, *274*, 25051-25060.

Lai, T. F., Li, B. Q., Qin, G. Z. & Tian, S. P. (2011). Oxidative damage involves in the inhibitory effect of nitric oxide on spore germination of *Penicillium expansum*. *Curr Microbiol*, *62*, 229-234.

Levine, A., Tenhaken, R., Dixon, R. & Lamb, C. (1994). H_2O_2 from the oxidative burst orchestrates the plant hypersensitive disease resistance response. *Cell*, *79*, 583-593.

Li, B. Q., Wang, W. H., Zong, Y. Y., Qin, G. Z. & Tian, S. P. (2012). Exploring pathogenic mechanisms of *Botrytis cinerea* secretome under different ambient pH based on comparative proteomic analysis. *J Proteome Res*, *11*, 4249-4260.

Li, H., Chen, Y., Zhang, Z. Q., Li, B. Q., Qin, G. Z. & Tian, S. P. (2018). Pathogenic mechanisms and control strategies of *Botrytis cinerea* causing post-harvest decay in fruits and vegetables. *Food Qual Saf, 2*, 111-119.

Li, H., He, C., Li, G. J., Zhang, Z. Q., Li, B. Q. & Tian, S. P. (2019). The modes of action of epsilon-polylysine (ε-PL) against *Botrytis cinerea* in jujube fruit. *Postharvest Biol Technol*, *147*, 1-9.

Li, H., Zhang, Z. Q., He, C., Qin, G. Z. & Tian, S. P. (2016). Comparative proteomics reveals the potential targets of BcNoxR, a putative regulatory subunit of NADPH oxidase of *Bortytis cinerea. Mol Plant Microbe Interact*, *29*, 990-1003.

Liu, J., Tian, S. P., Meng, X. H. & Xu, Y. (2007). Control effects of chitosan on postharvest diseases and physiological response of tomato fruit. *Postharvest Biol Technol*, *44*, 300-306.

Ma, D. Y., Ji, D. C., Zhang, Z. Q., Li, B. Q., Qin, G. Z., Xu, Y., Chen, T. & Tian, S. P. (2019). Efficacy of rapamycin in modulating autophagic activity of *Botrytis cinerea* for controlling gray mold. *Postharvest Biol Technol*, *150*, 158-165.

Malagnac, F., Lalucque, H., Lepere, G. & Silar, P. (2004). Two NADPH oxidase isoforms are required for sexual reproduction and ascospore germination in the filamentous fungus *Podospora anserina. Fungal Genet Biol*, *41*, 982-997.

Marschall, R. & Tudzynski, P. (2016). BcIqg1, a fungal IQGAP homolog, interacts with NADPH oxidase, MAP kinase and calcium signaling proteins and regulates virulence and development in *Botrytis cinerea. Mol Microbiol*, *101*, 281-298.

Marschall, R., Siegmund, U., Burbank, J. & Tudzynski, P. (2016). Update on Nox function, site of action and regulation in *Botrytis cinerea. Fungal Biol Biotechnol*, *3*, 8.

Minz Dub, A., Kokkelink, L., Tudzynski, B. & Sharon, A. (2013). Involvement of *Botrytis cinerea* small GTPases BcRAS1 and BcRAC in differentiation, virulence, and the cell cycle. *Eukaryot Cell*, *12*, 1609-1618.

Panday, A., Sahoo, M. K., Osorio, D. & Batra, S. (2015). Nadph oxidases: an overview from structure to innate immunity-associated pathologies. *Cell Mol Immunol*, *12*, 5-23.

Qin, G. Z., Liu, J., Li, B. Q., Cao, B. H. & Tian, S. P. (2011). Hydrogen peroxide acts on specific mitochondrial proteins to induce cell death of fungal pathogen revealed by proteomic analysis. *PLoS One*, *6*, e21945.

Qin, G. Z., Tian, S. P., Chan, Z. L. & Li, B. Q. (2007). Crucial role of antioxidant proteins and hydrolytic enzymes in pathogenicity of *Penicillium expansum* - Analysis based on proteomics approach. *Mol Cell Proteomics*, *6*, 425-438.

Reverter-Branchat, G., Cabiscol, E., Tamarit, J. & Ros, J. (2004). Oxidative damage to specific proteins in replicative and chronological-aged *Saccharomyces cerevisiae*. *J Biol Chem*, *279*, 31983-31989.

Roca, M. G., Weichert, M., Siegmund, U., Tudzynski, P. & Fleissner, A. (2012). Germling fusion via conidial anastomosis tubes in the grey mould *Botrytis cinerea* requires NADPH oxidase activity. *Fungal Biol*, *116*, 379-387.

Schürmann, J., Buttermann, D., Herrmann, A., Giesbert, S. & Tudzynski, P. (2013). Molecular characterization of the NADPH oxidase complex in the ergot fungus *Claviceps purpurea*: CpNox2 and CpPls1 are important for a balanced host-pathogen interaction. *Mol Plant Microbe Interact*, *26*, 1151-1164.

Scott, B. (2015). Conservation of fungal and animal nicotinamide adenine dinucleotide phosphate oxidase complexes. *Mol Microbiol*, *95*, 910-913.

Segmüller, N., Kokkelink, L., Giesbert, S., Odinius, D., van Kan, J. A. & Tudzynski, P. (2008). NADPH oxidases are involved in differentiation and pathogenicity in *Botrytis cinerea*. *Mol Plant Microbe Interact*, *21*, 808-819.

Siegmund, U., Heller, J., van Kan, J. A. & Tudzynski, P. (2013). The NADPH Oxidase Complexes in *Botrytis cinerea*: evidence for a close association with the ER and the tetraspanin Pls1. *PLoS One*, *8*, e55879.

Siegmund, U., Marschall, R. & Tudzynski, P. (2015). BcNoxD, a putative ER protein, is a new component of the NADPH oxidase complex in *Botrytis cinerea*. *Mol Microbiol*, *95*, 988-1005.

Takemoto, D., Kamakura, S., Saikia, S., Becker, Y., Wrenn, R., Tanaka, A., Sumimoto, H. & Scott, B. (2011). Polarity proteins Bem1 and

Cdc24 are components of the filamentous fungal NADPH oxidase complex. *Proc Natl Acad Sci USA*, *108*, 2861-2866.

Takemoto, D., Tanaka, A. & Scott, B. (2007). NADPH oxidases in fungi: Diverse roles of reactive oxygen species in fungal cellular differentiation. *Fungal Genet Biol*, *44*, 1065-1076.

Tanaka, A., Takemoto, D., Hyon, G. S., Park, P. & Scott, B. (2008). NoxA activation by the small GTPase RacA is required to maintain a mutualistic symbiotic association between *Epichloë festucae* and perennial ryegrass. *Mol Microbiol*, *68*, 1165-1178.

Tian, S. P., Qin, G. Z. & Li, B. Q. (2013). Reactive oxygen species involved in regulating fruit senescence and fungal pathogenicity. *Plant Mol Biol*, *82*, 593-602.

Torres, M. A., Dangl, J. L. & Jones, J. D. (2002). Arabidopsis gp91^{phox} homologues AtrbohD and AtrbohF are required for accumulation of reactive oxygen intermediates in the plant defense response. *Proc Natl Acad Sci USA*, *99*, 517-522.

Williamson, B., Tudzynski, B., Tudzynski, P. & van Kan, J. A. (2007). *Botrytis cinerea*: the cause of grey mould disease. *Mol Plant Pathol*, *8*, 561-580.

Wojtaszek, P. (1997). Oxidative burst: an early plant response to pathogen infection. *Biochem J*, *322*, 681-692.

Xu, X. B. & Tian, S. P. (2008). Salicylic acid alleviated pathogen-induced oxidative stress in harvested sweet cherry fruit. *Postharvest Biol Technol*, *49*, 379-385.

Zhang, Z. Q., Li, H., Qin, G. Z., He, C., Li, B. Q. & Tian, S. P. (2016). The MADS-box transcription factor Bcmads1 is required for growth, sclerotia production and pathogenicity of *Botrytis cinerea*. *Sci Rep*, *6*, 33901.

Zhang, Z. Q., Qin, G. Z., Li, B. Q. & Tian, S. P. (2014). Knocking out Bcsas1 in *Botrytis cinerea* impacts growth, development, and secretion of extracellular proteins, which decreases virulence. *Mol Plant Microbe Interact*, *27*, 590-600.

Zhang, Z. Q., Qin, G. Z., Li, B. Q. & Tian, S. P. (2015). Effect of cinnamic acid for controlling gray mold on table grape and its possible mechanisms of action. *Curr Microbiol*, *71*, 396-402.

Zhang, Z. Q., Tian, S. P., Zhu, Z., Xu, Yong. & Qin, G. Z. (2012). Effects of 1-methylcyclopropene (1-MCP) on ripening and resistance of jujube (*Zizyphus jujuba* cv. Huping) fruit against postharvest disease. *LWT-Food Sci Technol*, *45*, 13-19.

Zhu, Z. & Tian, S. P. (2012). Resistant responses of tomato fruit treated by exogenous methyl jasmonate to *Botrytis cinerea* infection. *Sci Hortic*, *142*, 38-43.

Contents of Earlier Volumes

Agricultural Research Updates. Volume 27

Agricultural Research Updates: Volume 26

Agricultural Research Updates, Volume 25

Agricultural Research Updates: Volume 24

Agricultural Research Updates: Volume 23

INDEX

A

B

C

D

E

O

P

R

S

T

U

V

W

X